REINCARNATION:

THE SCIENTIFIC THEORY

by

Michael A. Tewell, MA, JD

Copyright 2021

All rights Reserved

PUBLISHED 2021

REINCARNATION: THE SCIENTIFIC THEORY; copyright 2021 by Michael A. Tewell. All Rights Reserved. No part of this publication may be reproduced, stored, or transmitted in any form by any means, including but not limited to digital, electronic, mechanical, photocopying, recording or otherwise; or conveyed via the Internet or any website or other media without the prior, written permission of the author; except in the case of brief quotations embodied in news articles, reviews, academic research and books (electronic and printed).

Library of Congress Cataloging-in-Publication Data

Tewell, Michael A.

REINCARNATION: THE SCIENTIFIC THEORY/ Michael A. Tewell / a work of non-fiction which includes referenced citations and illustrations.

ISBN: 978-0-9600607-5-7 (paperback)

ISBN: 978-0-9600607-4-0 (ebook)

1. Reincarnation. 2. Evolution. 3. Genetics. 4. Immortality. 5. Individuality. 6. Neuroscience. 7. Quantum Mechanics. 8. Religion.

Cover Image: SelfPubBookCover.com / RLSather

Cover Design: SelfPubBookCover.com / RLSather

Inquiries should be addressed to the author: Michael A. Tewell;

P. O. Box 1314; Palm Harbor, Fl. 34683; or, at michaelatewell@gmail.com

PUBLISHED IN THE UNITED STATES OF AMERICA

CONTENTS

Contents..Page 3

Acknowledgements............................. Page 5

Introduction... Page 6

1. The Crossroads... .... Page 13

2. Fake News.. Page 23

3. PhiAlpha:The Seedof Life........................Page 37

4. Evolution: The Quiet Creator......................Page 62

5. Individuality: Child of Our Memories..........Page 82

6. PhiAlpha Activation...............................Page 93

7. Memory Organization............................Page 113

8. Memory Prioritization........................... Page 122

9. Memory Encoding..................................Page 134

10. Memory Emission............................. Page 143

11. Memory Reception.................................Page 150

12. Validation Testing...................................Page 161

13. Conclusions.. Page 174

14. PhiAlpha... Page 181

15. The Next Step...................................... Page 195

References.. Page 202

Illustrations.. Page 217

The Author.. Page 220

ACKNOWLEDGEMENTS

No one has sacrificed more as I spent hours researching and writing this book than my wonderful wife, Clementine "CC" Conde, and our own bright and shining star of a teenage son, Alex. Both had their own lives to live, yet they willingly put up with me as I wrestled this dragon of a book to completion. They are the best and I am forever grateful for their patience.

The author extends his appreciation to Creative Common licensees, who have allowed me to share their illustrations in this book. Certain parts of the theoretical model are technically worded, and the illustrations Creative Commons has allowed me to use are a valuable visual aid to help the reader understand the processes involved in PhiAlpha.

Finally, I am greatly indebted to my copyeditor, Sharon Kaplan of editingmatters1@gmail.com, for her extraordinary skill and professionalism. She has been a wizard with how she guided me painlessly through the treacherous landscape of manuscript editing. Without her steady hand, this book does not get done.

INTRODUCTION

Reincarnation is like a sleeping lion. You can live your life and not even know it's there. Then, one day, BAM! It hunts you down like a predator stalks its kill.

It happens when you least expect it. It can happen at any age. You start having dreams--very strange dreams.

These dreams, called "waking dreams" or "lucid dreams," occur just before you awaken. You watch as events unfold as though someone has turned on a motion picture projector inside your mind. Your mind is completely aware, yet a part of you still knows you are sleeping.

I began having these dreams after I turned twelve. I have had dreams of many different past lives. Most have to do with war, combat, killing—and dying.

Each time it has happened to me, it is a series of dreams over the course of weeks or months about the same events. To the minutest detail, each dream is exactly the same.

Also, these dreams are unusually graphic. Unlike normal dreams, in these "waking dreams" you are vividly aware of all your senses and all the sensory stimuli they experience. It's not like a dream at all. You believe you are actually experiencing something that really happened. You get the feeling you aren't dreaming but reliving an event.

My third set of "waking dreams" took me to Florida as a psychiatric patient at a VA hospital following WWII. I was really nuts—totally lost it. I was part of a Marine regiment tasked with attacking a Japanese-held island late during the war. The officers kept telling us there were very few Japanese on this little, unnamed island.

"It's supposed to be a cakewalk," my captain told us.

That "cakewalk" turned out to be Peliliu, I think. At least my present-life research matches the memories I have. That's what Reincarnation does to you. It gives you memories that you can NEVER forget because they are engrained in your brain as long-term memories.

Your past lives are "hard-wired" into your brain before you are born. You can run from them, but you can never hide.

After securing the beach, I volunteered for a solo recon of the jungle inland area. I was not being heroic; just bored, stinking hot, and going nuts from the sand flea bites. I needed a diversion.

Walking through the jungle, Thompson machine gun cradled on my right forearm, I heard birds calling and flying off trees nearby. Something had spooked them. It wasn't me. I was good at stealth.

Carefully making my way, I saw a trail. Following it, I came to a large banyan tree and walked around it.

There, sitting under another tree, was a Japanese soldier. For a few seconds, I just stood watching him. He was facing me, rifle by his side. writing something into a small journal or diary. He didn't know I was there. I had a bead on him, there was no stress.

Curious at how he'd react, I purposely coughed load enough for him to hear me. He looked up, and for a long second, we just stared at each other. Friend or foe? Living or dying? Maybe he wouldn't go for his gun. Maybe we could just smile at each other; be non-comrades-in-arms. I'd back off and we'd both live to see another day.

The journal and pencil fell to the ground. His arm slowly reached for the rifle. He must have known he had no chance. My Thompson was already pointed at him.

I squeezed the trigger without emotion. It was a cold-blooded reaction; programmed into my brain by intense Marine training to kill and not be killed.

But then something very strange happened—incredible yet horrific. As my bullets flew toward his chest, I saw his soul fly towards me. As my bullets punctured his chest, his soul punctured my own.

I went completely nuts. Sparing you, dear reader, the details, months passed in my dream. I end up a patient in a VA psych clinic in a VA hospital in Florida near a bay that was calming and very serene. The clinic was on the second floor of an old VA building. The first floor had a canteen used to sell coffee and snacks.

With a lot of therapy and time, I improved enough that I was given the job of mopping and waxing the first-floor canteen and hallway. I was very encouraged at my progress. I was healing. Then I went downstairs with my mop and bucket.

The first-floor canteen was built in a Bavarian-style architecture with dark-wooden rafters and designs painted in the ceiling. It was very pretty. Then I looked down at the floor I was to mop and wax.

The floor was made with tiles with the Nazi swastika! In a VA hospital in 1946, they have swatiskas on the damn floor! I was sure there was a conspiracy of Nazi's in our military and this was a private signal that the hospital staff were all Nazis!

But if the hospital staff is part of the conspiracy, who could I report it to? The "brass" in Washington? They were the ones who had sent me to this VA mental hospital to begin with. They had to be in on it.

It was up to me to fight this atrocity as best I could. I got my bucket full of lye and every other chemical I could think of and dumped it on the first-floor Nazi tiles. Then I went back upstairs like nothing happened.

The next day, the shit hit the fan, so to speak. I just played dumb. Didn't know a thing. Wasn't going to tell these Nazi's anything.

I started causing all kinds of hell in the ward. Not long afterward, I was discharged; more trouble than I was worth. Within months, I took my own life. That was somewhere around 1946-48. I was born in this life in September of 1950.

I began having that dream while working on my master's degree at Kent State University in the early months of 1979. I had no idea where that VA hospital was and had never heard of any VA hospitals in Florida in this life.

Then, in 1990, I started my career as an Assistant Public Defender in Pinellas County, Florida. One of my duties was to represent individuals subject to involuntary-inpatient psychiatric treatment (Baker Act) hearings. Once of these hospitals was at the Bay Pines VA hospital in Bay Pines, Florida, next to St. Petersburg.

I still remember my first visit there. Located next to beautiful Boca Ciega Bay, the Bay Pines VA psych clinic is on the second floor of an old building with a Bavarian-style interior, a first-floor canteen, and swastika-style tiles on the first floor—one area of which was seriously defaced by a very old stain of some kind.

At the time, I must admit to surrendering to a moment of some pride. I had never learned, in my most-recent past life, how effective my one-night as hospital-rebel had been before I committed suicide.

Reincarnation is real. It can be as hidden as your deepest memories, but when it awakens within you, your life will never be the same.

1. THE CROSSROADS

We stand at a crossroads. Our crossroads cannot be found on any map. It is a waypoint, not of space and place but of mind and time. Modern technological advances have brought humanity to a critical nexus where the gravitational pull of our ideological past, faith, and religious belief are racing headlong on a collision course with humanity's rush toward its future and scientific discovery. Nothing epitomizes this conflict between past and future, faith and fact, religious belief and scientific objectivity more than the issue of whether there is a life after death.

Is there life after death?

We all want to know the answer to that question. There may not be many questions each of us can agree on that matter in our lives, but we are all focused upon that one. That is the question this book will attempt to answer.

Priests, clergy, rabbis, and clerics have told our ancestors that there can be no life after death without religious faith. But let us ignore them for once. I am not talking about the afterlife we have been told exists by our world's many religions. What I am talking about, and what this book is about, is the question:

Does an afterlife exist independent of religious faith?

Is a nonreligious afterlife even possible? If it is, what are the processes by which such an afterlife occurs? None of the world's great religions attempt to answer this question scientifically. Their explanations of an afterlife remain frozen in time—written thousands of years ago in the Bible, the Torah, the Qur'an and other ancient texts.

Most of the world's religions profess the existence of an afterlife but none have ever been able or willing to describe the physical processes involved. Perhaps this is because religion is based upon subjective belief—not objective, verifiable, and reproducible truths required of true scientific inquiry.

Instead of scientific truth, our religious teachings enjoy the influence that thousands of years of indoctrination and ritual bestow upon it. But these teachings are not science; they are but faith. While faith has its power, without science there can be no certainty of truth. Without truth, a religion is mere illusion—a carnival with priests as mere clever showmen and humanity as a congregation of dumb puppets dancing, pulled by invisible strings. Without truth, faith evaporates in the air like moisture in the heat of the sun.

Over the course of the last 100 years, science has shown us what our religious faiths could not: increasingly more revealing pictures of the universe and our place within it. The Hubble Telescope and other satellites give us pictures of the cosmos almost to the very dawn of our universe. Geneticists have mapped both human and animal DNA and RNA. Researchers now program our DNA to resist disease and birth defects. Engineers now program machines (even automobiles) with artificial intelligence. Physicists use massive particle accelerators, such as the one at CERN, to discover ever smaller and more fundamental elements like the Higgs boson and bring us ever closer to replicating the Big Bang.

If we are capable of all this, is not our technology now advanced enough to begin the process of unmasking the greatest mystery of all—the afterlife? If we can spend billions of dollars searching for intelligent life on other planets, we can begin to look scientifically at whether there is life after death.

Does an afterlife exist, or not? Surely, our technologically advanced civilization is mature enough to put the afterlife to the test. But can our science now show us the truth of what lies beyond "the pale frontier" we call death? Is it not time to begin the study of life beyond the grave?

More to the point, are we now ready to take the bold next step in our intellectual and spiritual enlightenment, and face the challenge the following question poses to each of us?

Humanity is ready for this journey. In fact, humanity is desperate for an opportunity to free itself from the intellectual chains of thousand-year-old ideologies that have less to do with individual spiritual development than the manipulation of societies through mind control of the masses.

So, let us turn the question inside out. Instead of whether an afterlife exists, let us pose the following question instead:

Does an afterlife exist dependent solely upon natural laws?

This book presents a biologically based theoretical model of life after death independent of religious faith. This paper explores the possibility that an afterlife does exist—not as a religious expression—but as a purely biological one; dependent solely upon natural laws.

We do not present proven fact. Presented here is still only a theoretical model. However, this theory is supported by modern scientific research. Each of the biological processes necessary for our theoretical model to work is validated by published scientific research, each cited herein for easy reference.

What this means is that the biological processes necessary for an afterlife to exist are present in the human body. Until now, they have remained hidden, like pieces of a puzzle waiting for someone to put them together. The answer to this puzzle begins with a simple question:

What is the objective purpose of life?

We understand there are many religious and philosophical perspectives regarding this question. However, we do not concern ourselves with either religion or philosophy here. Here, those issues no longer matter. Here, we are going back to square one, before all the religious hyperbole and philosophical narcissism. Here, there is only one issue—life itself.

From the very moment life on Earth began as single-celled organisms with cilia reaching out into the external environment for stimuli, existence has been one with the quest for knowledge.

Billions of years later, cilia have become nerve endings reaching out in skin-covered limbs, and stimuli is no longer other bacteria but the universe itself. Nevertheless, the concepts and precepts upon which life began remains with us today.

We, the human ruling class of the modern world, cushion our daily existence with the superficial luxuries of technology. We stand upon the summit of the pyramid of biological evolution. Strip away our science, remove our engineering, and we share the same priority that dominates the existence of every other species on the planet. We wish to survive. We need to survive. Survival is the driving force that propels all we do. Its premise is inlaid within every motivation and every action we take.

Survival for us, like for every species that came before, depends upon one thing—information. We worship information. We spend our lives in the pursuit, collection, preservation, and sharing of information. All human life—from the very sensory inputs of our five senses to the love of history, literature, science, the arts, construction, sports, music, religion, philosophy, law, and politics—is ultimately intimately intertwined with information. The context may change, but the urge to know remains fixed and unwavering.

Why?

Information is nothing more than stimuli evolution has programmed us to rely upon for our individual survival and experiential growth. The information we receive during a lifetime is stored in our neurons as memories. Our brain is a vast memory bank of everything we have experienced—not only in this life but from past lives as well. What is the ultimate application of all this information? What is being created or conserved?

We label this creation "individuality," which every human possesses as a construct of one's consciousness and memories. Every life is unique. Every life has unique episodic sequences of events. These event sequences are stored in our neurons as memories. Every time we consider our past, our memories flow like water over the emotional waterfall of our limbic system. This waterfall of memories through the limbic system constructs our identities, our individualities. This information is shaped by our own uniquely evolving personalities and then formatted as neural and synaptic memories which are in chemical communication with our DNA and RNA codes by evolutionary processes.

Does evolution find individuality of information the main point of human cerebral growth? If so, and if individuality is so important, evolution would need to protect its existence. What biochemical processes would evolution likely create to protect its most important creation—individuality?

This book presents the theoretical model of "PhiAlpha." This model consists of a complete, multisystemic, biological process which preserves our personal, unique individualities from one lifetime to another. The reader can relax about how difficult this book is to understand. It is very easy. It was written by and for the nonscientist. PhiAlpha applies to everyone, so it is important that everyone can easily grasp and digest its processes—nonscientists as well as scientists.

This book walks the reader through the scientific concepts of PhiAlpha in layman's terms as much as possible. If the reader is able to understand the concepts presented in this chapter, the rest of the book should be just as easy to read and understand.

I appreciate that many will find this theory a threat to their religious upbringing. These fears are unnecessary. The universe created whatever we experience. Reincarnation is just another natural process of that experience—just as life and death. In fact, Reincarnation should take away the fear of death—and give life new meaning. Will it change the world?

This is a science book written for the average person. It is also a New Age book that explores the edges of our spirituality and religion in this uber-modern age.

Does an afterlife exist—or not? The world's religions have had thousands of years to prove its existence and shown us nothing of objective, reliable consequence. Can science do any better?

2. FAKE NEWS

Even as Nordic Vikings routinely sailed between Newfoundland, Nova Scotia, and other areas of the New World, Christian Europeans were still being brainwashed by their priests and monarchs that the world was flat. The Christian church was threatened by the existence of any New World that might challenge its authority and control over the descendants of the Greco-Roman empire.

It took a leap of faith by Christopher Columbus to put that "fake news" to the test. As a result, Columbus's trek to the New World broke the chains of intellectual imprisonment of an entire civilization and opened a new continent for exploration and discovery.

That was hundreds of years ago, but those same chains still exist today. Those same chains still limit the intellectual freedom of discovery and enlightenment in people all over the world when it comes to thinking about death and what lies beyond.

We die. We die.
Original sin is why.
Religion never lies.
We have always died.
Religion rules the afterlife.
Never ask:
Why?

For thousands of years, this has been the human mantra. Reinforced by countless sermons and rituals imposed upon us by our own religious faiths; this myth has prevented even the most adventurous of human minds from asking the biggest question of all:

Evolution controls our life; does it also control our death?

For thousands of years, our ancestors accepted whatever the religious elites told them to believe about death and whatever follows afterward. Even today, a review of the scientific literature reveals an astonishing truth: a life-after-death s cience does not yet exist.

But there are questions that logic compels us to answer:

1. Why do we die?

The law of conservation of energy states that energy cannot be lost. Energy can only change its form from one state to another through transitions consistent with natural laws.

So why do we die?

2. What happens to our energy after we die?

If our energy cannot be lost but only transformed from one state to another, what is that state?

3. How is our energy transformed when we die and where does it go? In other words, what are the specific physical processes involved in the transit of energy from our dying bodies to…whatever exists beyond?

These are the questions that any life-after-death science would ask. But, aside from anecdotal reporting from curious emergency room physicians who have witnessed events that have no apparent explanation, scientists have been unwilling to study these issues. As a result, there is no life-after-death science.

The question begs: Why not?

While there have been countless writings on the issue of an afterlife, they have all been religious in nature. The Egyptian Book of the Dead, Hindu Shruthi, Islamic Qu'ran and Christian Bible each have their own explanations of god-hood and the afterlife, but the premise of each is the same. Each writing is a subjective account of a personal relationship with a super-being who vaguely describes an afterlife and the behavior required of anyone seeking to be granted the gift of a life after death.

Only in the leading Hebrew religious texts (Torah and Talmud) do we find an overt reluctance to commit to a set belief structure on an afterlife or *olam ha-ba* (jewishvirtuallibrary.org). It is suggested there that this reluctance was forged out of the Israelites' time in Egypt, when they may have developed an antipathy toward the Egyptian obsession with death.

History teaches us the causal effects of our blind acceptance of theocratic indoctrination. We know what that means. Religious fervor has killed more humans than any other causal factor in history.

The list is as long as it is tragic: ritual human and animal sacrifices dating back thousands of years, Incan and Mayan human sacrifices, the "Christian" crusades, Islamic jihads, the "Christian" Inquisition, the French Wars of Religion, the Holocaust, Ku Klux Klan murders in the American South, and genocide of native North Americans are just a small sampling of the violence humans have inflicted upon their brothers and sisters in the name of religion (and you can see that the predominant offender is the Christian religion, which often is anything but Christ-like).

The beast of violence in the name of religion continues today. Radical interpretations of the Qu'ran spread like cancer throughout the Middle East and the rest of the world, creating a demented ideology of death for misguided fanatics convinced that it is holy to murder innocent men, women, and children.

The beast of religious intolerance and hate spawn death and destruction even today: abortion clinic bombings (Christianity); the Boston Marathon bombing (Islam); the Israeli-Arab conflict is a continuing struggle between religious-based cultures and the nations of the Middle East (Islam and Judaism); and the Fort Hood and Orlando Pulse nightclub shootings (Islam)—to name just a few. No religion is immune; all have sinned in the persecution of others.

Religious indoctrination attacks the mind. A person who was once calm and reasonable before religious training becomes an intolerant fanatic afterward. Religious orthodoxy requires absolute obedience, and that obedience comes at a price—an ever-increasing loss of one's freedom of individual expression.

This loss of freedom of expression is only the beginning, however. Once one's individuality is lost, so too are one's ideals—even basic morality. Eventually, a mob mentality takes over. The result is no longer an individual identity but an inhuman machine willing to kill a complete stranger in the name of religion.

The effect of this religious-created blindness has had its pernicious influence upon scientific expression as well. Over the thousands and thousands of years of human recorded history, we see many expressions of religiously defined interest in the afterlife—but none regarding any nonreligious or scientific alternative beliefs. It is as though when it comes to the issue of life after death, humans can only think in terms of religious explanations. We are blinded by our religious upbringing. It is this very blindness that has kept humanity from looking for an alternative explanation to life after death all these centuries.

The myths and mysteries of life after death have come to us through cave art, fables, clan tales, prophets, priests, martyrs, and scribes since the dawn of history. Human history is, itself, a chronological recitation of our ancestral dependence upon the crutch of religious faith to face each new day. We have all been affected by it. Our perceptions of reality have been so intertwined with religious iconography that we find it difficult to see the universe for what it truly is. In our blind rush toward subjectively perceived emotional safety, we closed our eyes to that which objectively confronts us—the objective universe.

The veil, woven by our ancestors, eventually becomes our own. The deception is planted at infancy by those we trust—our parents. The deception is reinforced throughout life, every season of every year, through community festivals that serve to reinforce the collective faith in the deception—as well as to create cognitive dissonance within any individual whose mind seeks to break free of the deception to search for an independent reality. Community pressure is brought to bear upon any individual who, in confession or to family, admits to being "tempted" to break away.

For most, there soon comes a time when the indoctrination is so overwhelming, the individual pursuit for truth inherent within every soul is forsaken as the price paid for community approval and acceptance. There is then no longer a need for external reinforcement. The deception has taken root within the psyche and one's own mind becomes the principle enforcer of the deception. The objective universe no longer exists. In its place is the fictional universe of theocratic subjectivity.

That is the case for most us. Most of us want, even need, to be liked by our community. Conformity is easier than being cast out as a heretic—or worse—for rejecting the norms of our religious traditions. The psychological pressures to conform a worldview of one's community is often too great to resist for long. Resistance is, in the end, futile.

One's own ideals surrender to the will of one's less curious and more indoctrinated peers.

As independent entities of individual consciousnesses we have a right to pursue our own inner journeys of revelation independent of others. But these rights come with the duty to respect one's own intellectual independence by continuing education of the self and open-mindedness to the opposing views of others.

But religion is not the same as spirituality. We should not confuse the two. Religion is merely how we humans organize ourselves within our communities. Religion is impersonal, public, outward looking, and temporal. On the other hand, spirituality is our personal experience with the universe and our existence within it. Spirituality is personal, private, inward looking, and timeless.

Too often, however, we get the two confused or we sacrifice our spiritual integrity for the comfort (however temporary and superficial) of acquiescing to the social norms inherent in religious beliefs of our own communities. But by so doing, we become complicit in our own phenomenological ignorance. We allow others to tell us what to think and how to express ourselves in actions without any significant resistance. Worse, some have witnessed violence and hate in the name of religious faith without the will to stand opposed. They have become complicit.

Nevertheless, there is hope. Time and events propel us to our shared destiny independent of our finite intentions and expectations. Flickers of light within the gloom of our private intellectual caves have been observed regarding the question of an afterlife. The flickers create shadows. The shadows reveal random activity that begins trending toward a burgeoning realization of inevitable significance.

Thus, we witness the dawn of a new age—the birth of a life-after-death" science.

Toward that end we present what few articles of scientific research on the broad scope of near-death experiences we found pertinent to our purposes. While the papers presented herein cannot be classified as an "afterlife science," they do constitute a representative sampling of what ancillary research exists in related fields:

Peter Craffert, PhD, is a professor in the Department of Biblical and Ancient Studies at the University of South Africa. His 2015 article reviewing the last 40 years of research into nonlocal consciousness argues that research of out-of-body and near-death experience (NDE) have failed to prove that consciousness can exist outside the human brain. Accounts of NDE are not evidence without objective proof. Further, he argues, testimonials of personal perceptions of NDE are anecdotal at best and at worst, hallucinatory.

Craffert's concerns remind us that extraordinary claims require extraordinary evidence, and radical views require radical evidence of the highest quality.[1]

However, a scientist and two medical doctors have reported observations of phenomena that appear to avoid the errors Craffert warns us against and strongly suggest he is mistaken.

Born in Inner Mongolia, Jimo Borjigin received her doctorate at the prestigious Johns Hopkins University in Baltimore, Maryland, and has been an associate professor at the University of Michigan Medical School. Borjigin's work has earned her both the Life Sciences Research Foundation Fellowship and the John Merck Scholars Award.

Studying rats, Borjigin discovered their brains exhibited surges of highly coherent, synchronous gamma oscillations that occurred within the first 30 seconds of cardiac arrest supporting heightened consciousness at death. This high-frequency burst consisted of a striking increase in anterior-posterior connectivity in both theta and alpha wavelengths and exceeded any levels produced during normal, healthy waking states.[2]

Before Lakhmir S. Chawla, MD, was appointed chief medical officer at La Pharmaceutical, he was a practicing anesthesiologist and taught medicine as an associate professor at The George Washington University School of Medicine and Health Sciences in Washington, D.C.

Chawla's research involved human patients at end of life, which he observed in his work as a practicing anesthesiologist at the GWU hospital. Unlike Craffert, Chawla witnessed the process of death firsthand many times. Chawla reports that of the approximately 100 human subjects whose death he witnessed, 80 percent exhibited an unexplained EEG spike at death.[3]

While conceding no one knows what causes these EEG spikes, Chawla concludes this EEG spike observed in so many patients, at death, cannot be a random event and suggests the spike is a telltale signal of a physiological process, the cause of which, remains a mystery.

James H. Lake, MD is a board-certified psychiatrist who has published research in peer-reviewed journals on integrative medicine and psychiatry with emphasis on the role of consciousness and intentionality in human healing. He, too, has observed sudden EEG spikes at near-death events and wonders why they happen.

Lake acknowledges he does not know the cause of, or the reason for, these obviously nonrandom, near-death EEG spikes. To explore this mystery, he proposes what he terms a "testable neural model" based upon his research. He sees the near-death experience as a unique event consisting of the encoding and decoding of information in cortical and subcortical brain regions. He proposes a longitudinal study using sophisticated EEG and brain imaging techniques with advanced data analysis to further explore these EEG surges to determine their origin and purpose.[4]

That is all there is—for now. To put the issue in proper historical perspective, consider this: Regarding life after death, we are where Sir Isaac Newton stood when he witnessed the apple first fall from the tree. All Newton knew was what he observed. Without a theoretical model to go by, he had no hope of understanding the force causing the apple to fall.

That is where we are now. We stand in Newton's shoes, watching flesh die without a clue what force or forces control it, why it happens, or what happens to the energies released afterward. We observe the superficial processes of physiological death. But we have no idea how those processes relate to the natural laws governing the physical processes of the universe.

Today, however, we stand at the doorway of such a discovery. Behind us are all the myths, fables, and mystical hocus-pocus with which we blinded ourselves over the centuries of intellectual ignorance and fear. Waiting for us ahead is the mystery of life in a causal universe that science is only now beginning to explore. Who can say what we will find?

3. PHIALPHA: THE SEED OF LIFE

Simple observation teaches us that life is a phenomenon dependent upon the interaction of multiple processes occurring both within and outside the body. If life after death exists, it too must be dependent upon the interaction of separate processes from multiple biological functions for it to occur. In other words, if we live in a physical universe ruled by physical natural laws, then anything existing within this universe must obey those laws—even the afterlife.

The phenomenon of life after death has become so identified with the nonscientific and the supernatural that to objectively study the process or processes involved in the death, we must distinguish them from the phenomenon itself. Without some means of differentiating the two, we lose focus upon both. This lack of focus and differentiation interferes with scientific analysis by blurring the edges between both until we begin to confuse one for the other.

This problem can easily be avoided by giving the biological process we are describing a name. Such a name will need to respect the scientific nature of its existence and properly symbolize its theoretical function and purpose.

Therefore, the biological process that we believe create life after death shall be termed "PhiAlpha" (ΦA). We use the Greek alphabet because of its traditional use in science. We use "Phi" and "Alpha" because of their symbolic meanings within the Greek alphabet.

The Greek letter phi (Φ) is a combination of two other Greek letters: iota (I) and omicron (O). The circle of omicron symbolizes how our actions in life define the space and meaning of our existence. Who we are as individuals is defined or encircled by the circumference of our thoughts and deeds. The greater the impact of one's life, the farther out from the center of our being extends the radius of the circle. The straight line of iota symbolizes the union of two natural laws:

Natural Law

(1) **The conservation of energy—the total energy of an isolated system remains constant.**

(2) **The second law of thermodynamics—if the physical processes are irreversible, the combined entropy of the system and the environment must always increase.**

We end our symbol of PhiAlpha with the letter alpha (A) because the result of the unified processes involved in our model create a new physical life for the individual. Therefore, when we wish to identify the theoretical model of life after death, we shall use the term "PhiAlpha."

What is PhiAlpha?

PhiAlpha is an evolutionary process of conserving individuality from one life to another. By using multiple physiological and biochemical processes that conform to fundamental laws of nature, individuality is transferred from the dying into the living. It is a harmonic transfer of information obeying classical and quantum mechanical principles.

We define existence itself as a state of being of coherent energy whose main purpose appears to be the pursuit of ever more complex expressions of individuality. We exist eternally as coherent energy, but that coherent energy experiences transformations in the form of changes of phased states of matter (energy expression) that we have come to know as lifetimes.

Therefore, we define life as an intermediate state of existence phased in biological form. One of the purposes of life is the acquisition, reflection upon, and sharing of ever more complex information that enhances the expression of ever more complex forms of individuality.

The origin of PhiAlpha

1. The universe is a relatively closed system of energy and matter. This is the living environment

   existing in a state of increasing entropy (uncertainty).

2. The living environment creates information in the form of sensory stimuli.

3. Into this experimental tableau are placed living creatures. These creatures must use the sensory organs they are provided with to navigate the perils of their living environment. The purpose is simple—survival.

4. Survival of the fittest (natural selection) becomes an end unto itself—evolution. Some believe evolution has no purpose other than survival. They are wrong. Species survival is not the purpose of evolution. Survival is a simple catalyst. Survival is the natural catalyst to force individual living beings to make choices. From one-celled bacteria to humans, all living things throughout time have been forced to make choices necessary to survive. Some succeed and live to pass on their DNA to future generations. Others make wrong choices. They fail to survive.

Sounds like the algorithm for a new role-playing video game doesn't it? But it isn't.

It is real life.

All the choices individual creatures make are necessary for natural selection to function. Choose correctly, and you survive to procreate and improve the gene pool. Choose incorrectly, and you die. Your genes are lost forever.

One cannot study how life evolved from microscopic amoeba to the highly complex human beings we have become without sensing and appreciating evolution's logical progression. There is a definite purpose to evolution. What purpose—beyond ever-increasing complexity of individual expression—we have yet to understand.

Some academicians argue evolution has no purpose beyond survivability. This is akin to saying the universe has no purpose. They lack the vision to see beyond their narrow focus of scientific expertise. They fail to realize evolution is greater than the sum of its parts. They are so flush with excitement over what they have learned that they are blind to the reality that what they know is comparatively insignificant to what remains hidden.

We need more scientists with the humility to appreciate the unknown and less who waste their careers congratulating themselves on what comparatively little they have discovered to date.

The bottom line is that for all we think we know about evolution, it remains an enigma.

We know what evolution has done to us so far; but we have no clue what it will be doing to us in the future. No physical process exists without a purpose. We know nothing about evolution's origin. Where did it come from? How was it created: organically or artificially? How did it become part of our biological, cellular makeup?

Until we can answer those questions, we cannot say we have a clue on the more pressing question: Where is evolution leading us—and why?

5. Evolution creates ever-increasing complexity of information and the biological processes necessary to adapt to it for survival. As the environment increases in complexity, so do our cognitive abilities. One begins to see both environment and evolution as the opposing ends of a rubber band. When one pulls, the other follows. Among the many adaptations that evolution adds to our cognitive abilities, the most important becomes that of memory.

6. Over time, the information stored in human memory becomes the framework within which the consequences of the choices we make become our emotions. This emotional memory is a unique blend of information plus emotion—our individuality.

7. Natural laws prevent the destruction of energy and information.

8. Human death interferes with information conservation (survival of individuality).

9. As evolution must obey natural laws, biological adaptations would be constructed through evolution to ensure that individuality is conserved as unique information.

10. PhiAlpha is what we call the biological adaptations evolution would most likely choose.

The following is an absolute truth:

"The laws of thermodynamics are special laws that sit above the ordinary laws of nature as laws about laws or laws upon which the others depend. It can be successfully shown that without the first and second laws (of thermodynamics) which express symmetric properties of the world, there could be no other laws at all."[5]

Thermodynamics began as a study of how steam behaves under differential pressures and temperatures over time inside confined spaces such as steam engines. However, scientists discovered that the principles governing heat and pressure in such confined spaces are merely applications of fundamental laws governing any process that cause a change of state of being of the particle or particles involved. There are four separate laws of thermodynamics. We concern ourselves with only the first two.[6]

The first law of thermodynamics

"All real-world processes involve transformations of energy, and the total amount of energy is always conserved (stays the same). The universe is a continuum of energy which remains the same regardless of transformation. As to energy, time is symmetrical. The same energy exists for all time, simply changing states over time."[7]

The second law of thermodynamics

"In all natural processes, the total entropy of a system plus its environment cannot decrease. It can remain constant for a reversible process but must always increase for an irreversible one. The universe acts spontaneously to minimize potential outcomes of any process. Therefore, the universe always seeks to maximize disorder (entropy). As to entropy, time is asymmetrical. Therefore, the Second Law of Termodynamics creates a measure of the transformation of energy over time."[8]

The concepts of entropy and information are often misunderstood. Entropy is a term used to describe the relative disorder existing in a closed system. Some prefer to define "entropy" as "disorder." However, we prefer to use the term "uncertainty" instead.

Why?

The answer to this question is critical to our inquiry. The term "disorder" is relatively vague. The term "disorder" can mean something severe, like "chaos" or something minor, like "messy." Neither terms—"chaos" nor "messy"—have the necessary objectivity science requires. Neither can they be objectively measured or analyzed. They are both too subjective to have scientific consequence.

"Uncertainty," however, implies a lack of something objective—information. Information can be measured, analyzed, and transformed from one state to another. Information is the difference between uncertainty and certainty. Information is the balance between uncertainty and knowledge.

To illustrate, consider this thought experiment:

You are walking alone in the desert for hours. You walk until you cannot walk any longer. Stopping to rest and refresh yourself, you look around. As far as your eyes can see, there is only the sunbaked sand of the desert.

On a whim, you reach down and pick up a handful of sand. It is hot to the touch, but at least you know where the sand is. The uncertainty or entropy is low. You know where it is. It is in your hand. You feel it. You know it is hot, dry, and hard. You possess all of the information you need to know what the sand is.

Now, you suddenly throw the fistful of sand into the air. The second the sand leaves your hand, the uncertainty or entropy of the sand increases 100 percent. You no longer possess it. Each grain now possesses its own trajectory, independent of the others. You do not know how long each grain will be in the air or where each will land. This uncertainty increases over time the longer each grain flies in the air.

But here, as everywhere else in nature, there is balance. While uncertainty increases over time, so too does information. So, you watch and observe the grains of sand as they arc their way above the desert floor until they land. Your sense of sight gives you information on where and how far each granule of sand travels and where each one lands. While uncertainty gradually increases as each granule travels through the air, this uncertainty is continuously being resolved through the information from your sense of sight. A balance is struck. What entropy would conceal, information both reveals and conserves.

Our universe, being a closed system, is subject to the laws of thermodynamics like any other. At the time of the Big Bang, entropy (uncertainty) was relatively low. Over time, however, the level of entropy has increased. This change in the degree of entropy can be measured. Information, over time, creates measurements of change to our state of being. This proposition leads us to the surprising but logically inevitable conclusion that all knowledge is reducible to classical mechanical analysis.

Since the second law of thermodynamics provides a measurement of change over time, scientists have concluded that this second law governs the conservation of information. Information is the result of any observation or measurement. One may even conclude information equals existence and vice versa.

When any object undergoes a transformation, it creates information about its change of state during its transformation. All objects, including biological lifeforms (humans) create information describing the relative state of being during their existence. Information is created whenever any object undergoes transformation within a system as it is being transformed.

In 2013, a research team from the People's Republic of China, led by Dr. Baocheng Zhang and Qing-yu Cai, PhD, published a paper on information conservation that is a beautifully articulated presentation of the principles involved. Their conclusion speaks not only to classical mechanics, but to the realm of quantum mechanics as well. Both Zhang and Cai are professors of physics at Hong Kong Polytechnic University. Their report states:

"Information is physical, it cannot simply disappear in **any** [emphasis added] physical process. This basic principle of information science constitutes one of the most important elements for the very formulation of our daily life and our understanding of the universe." [9]

The reader should not ignore the word "any" in the above quotation's first sentence. Zhang and his team are trying to explain the broad implications of information conservation theory.

When they state information is conserved in "any" physical process, they mean just that—any. That includes the physiological lifecycle of humans.

The human physiological lifecycle is a physical process just like any other, and so it must conform to the law of information conservation. The information within the universe cannot be lost or destroyed. Theoretically, if information could be lost or destroyed, then the Big Bang itself could be undone. That is an impossible paradox.

The father-son team of Ryszard Horodecki, PhD, and Michal Horodecki, PhD, are just two of an entire family of physicists from the University of Gdansk, Poland. Michal is internationally recognized for his discovery of quantum-state merging and for co-discovery of the Peres-Horodecki criterion for testing whether two quantum states have become entangled. Together, the two Horodeckis reformulated the traditional second law of thermodynamics to better conform to quantum theory. Their redefinition of the second law of thermodynamics according to quantum theory is as follows:

"In quantum mechanics, one cannot either delete or clone [quantum information]. The Second Law holds: entropy cannot be decreased in a closed system. Again, the Second Law can be stated

in a stronger way: that eigenvalues of density matrix do not change under quantum evolution. And this is actually the information conservation principle."[10]

From this redefinition of the second law came two principles of information conservation:

(1) The No-Deleting Principle

"In a closed system, one cannot destroy quantum information. Quantum information can only be moved from one sub-space location to another. In a closed system, even orthogonal [opposite] states cannot be deleted alone, given unitary evolution. Once entropy or information has increased, it cannot be decreased—it can only be transferred to a separate location."[11]

(2) The No-Cloning Principle

"Quantum information cannot be 'leaked-out' while keeping the original information intact. This is true for both open and closed systems, as well. This principle is already provable using unitary evolution dynamics. Once information is increased, it cannot return to its original state."[12]   In other words, quantum information cannot be duplicated. Therefore, information is unique unto itself.

A physicist, Marco Roncoglio's field of expertise is quantum mechanics and theoretical physics. He is a past visiting fellow at the University of Munich's Department of Physics. His paper, "On the Conservation of Information in Quantum Physics," introduced the concept of coherent information being conserved in a quantum state. According to Roncoglio:

"...[P]ure states contain more coherent information than mixed states, as the missing information has been converted into correlations with the environment. We will find that the coherent information, associated to genuinely quantum phenomena, is indeed conserved under unitary processes."[13]

Roncoglia's "correlations" sound very much like the neurally encoded memories we observe in the human brain. Put simply, what we experience as neutrally encoded stimuli is Roncoglia's "pure information."

This information is "pure" because it is raw sensory input. There is no emotional bias affecting its coherence. Through contemplation, analysis, or simply impulse, we mix this "pure" sensory input with the emotional resonance created from memories from past experiences. In this process, some of the pure data is "lost" to our consciousness but never really lost. While lost to our conscious awareness, the original, sensory information remains conserved under quantum processes.

That is how information conservation becomes important to our human existence. It places the information we use to create our own unique information—individuality—within the framework of quantum mechanics and the second law of thermodynamics.

PhiAlpha is based upon two fundamental principles:

(1) Evolutionary adaptations of the human brain created the process of creating unique information we call "individuality." This individuality is unique information created through the interaction of new stimuli (pure information) with emotionally loaded memories of past experiences (mixed information) and;

(2) The quantum mechanical application of the second law of thermodynamics upon this unique information we call "individuality" compelled the evolutionary adaptations that created PhiAlpha—the biological processes that conserve individuality from one life to another.

Objections to PhiAlpha

There are three basic objections to the PhiAlpha theoretical model. The first objection is that PhiAlpha is not testable and that until PhiAlpha is testable, it must be regarded as nonscience. People who believe this are mistaken. One aspect of the PhiAlpha model has not only been tested, it has already been proven. It is as follows:

Before her tragic death on her 27th birthday in 2009, Jharana Rani Samal, was one of India's brightest young physicists. She and her research team conducted an experimental test of the information conservation's no-deleting theorem (her team called it the "no-hiding" theorem) by using nuclear magnetic resonance and quantum state randomization of a qubit to show that the missing information can be fully recovered. This test proved that if any physical process results in the apparent loss of information, the missing information is not lost but merely relocated.[14]

In her honor, co-researchers Arun Pati and Anil Kumar had her work published posthumously in 2011. Pati has been professor of quantum information at one of India's leading physics research labs. In 2008, he co-edited *Quantum Aspects of Life*, which included the singular honor of having its forward written by none other than England's world-renown physicist-mathematician Sir Roger Penrose. Pati's articles have earned publication in such noted science journals as *Nature* and *Science*, and he is an elected member of India's National Academy of Sciences.

True, none of these researchers understood the significance of their discovery at the time. But no one has ever before considered the application of information conservation principles upon the human memories that create our unique identities.

The second objection or challenge to PhiAlpha is based upon a perceived lack of evidence. Some claim PhiAlpha does not exist because no evidence of its existence is observed. There is a false logic at work in this supposition, however. For example, there is no evidence that aliens exist on other planets. Does that lack of evidence convince us not to search? No, of course not.

We do not need tangible physical evidence of the presence of alien life on other planets because we believe they do exist. The weight of the logical inferences we have drawn from our life experience compels our faith in such a conclusion. Such is the case with a biological process for life after death. The weight of logical inference based upon our life experience forces the conclusion upon us.

The third challenge arises out of the problem of transferring information from a dying body into one that is living, one that is about to be born. Without a transport mechanism in place to secure our memories and transfer them from our dying body to an unborn one, this is all just a waste of time—isn't it?

This is true.

Furthermore, wouldn't this transport mechanism have to be a quantum mechanical one at that?

Yes, this is also true.

Then are we not wasting our time? It's impossible to find such a mechanism, isn't it?

Wrong.

The transport mechanism we need already exists. The transport mechanism we are looking for is found within every cell of our bodies—biophotons.

The team of Siyuan Shi, Prem Kumar, PhD, and K.F. Lee proved the biophotonic transmission of information (energy) in 2017. Kumar has served as professor of physics at Northwestern University's Center for Photonic Communication and Computing. Together, they generated a polarization-entangled two-photon state in an ensemble of fluorophones. According to their report:

"Recent development of spectroscopic techniques based on quantum states of light can precipitate many breakthroughs in observing and controlling light-matter interactions in biological materials on a fundamental quantum level."[15]

Biophoton packets can be entangled to transmit coherent information almost instantaneously and, seemingly, without limit. Researcher Tristan Tentrup, a promising Ph.D. candidate from the University of Twente in The Netherlands, and his team claim that "encoding information in the position of single photons has no known limits. ..."[16]

Even our military has recognized the importance of using photons as a carrier of information.

In a rare unclassified research report, the United States Defense Advanced Research Projects Agency (DARPA) describes the photon as "... a fundamental carrier of information, possessing numerous information carrying degrees of freedom including, frequency, phase, arrival time, orbital and angular momentum, linear momentum, entanglement, etc."[17]

The purpose of the DARPA program (dubbed InPho) is to pursue the fundamental science and applications of photonic communications (biophotonic and non-biophotonic) in communications and imagery transfer.[18]

DARPA's application of photonic communication for military communications and imagery provides a near-perfect match with our PhiaAlpha model of neural-memory-information processing within our own human brain at death. Later in this book, we apply this knowledge of biophotonic transportation of information to create our theoretical model of life after death.

We conclude that the process humans have termed "death" is not an end of life but a transformation of the coherent energy created by biological life when the physical body decays. Without PhiAlpha, the energy of the biological lifeform would lose coherence at death, causing the dissipation of consciousness along with all the knowledge and memories collected during life. PhiAlpha creates a lifeboat of energy coherence for memories to remain coherent between the end of biological life of the old host and the fetal beginning of the new.

PhiAlpha and Consciousness

Does this mean consciousness is being transferred from one physical body to another?

No.

PhiAlpha is separate from consciousness. The study of PhiAlpha may or may not yield comparative information as to consciousness, but its purpose is distinct from that of consciousness. PhiAlpha is a process intended to conserve individuality as an expression of ever more complex information—which may be an intended or unintended consequence of consciousness. But they are not the same.

Consciousness, we believe, requires self-awareness—which is more than simple identity. Consciousness appears to be different than PhiAlpha, which is merely an individual identity matrix built upon biochemical constructs called memories and the mechanisms evolution created to preserve their existence.

4. EVOLUTION: THE QUIET CREATOR

There is but one interface between life as we know it and the laws of conservation of energy and information. We call this interface genetics or evolution. While DNA and RNA control and predetermine form and context for all life on Earth, these genetic codes must obey the same fundamental laws of the universe as any other physical process.

Though we do not yet know how evolution began or what (or who) created it, we do know evolution is a process created from within the fundamental laws of the universe. Therefore, its function and biology must act in accord with those universal laws. We are only now beginning to understand genetics. The relationship between genetic evolution and the forces and elements that combine to create life remain unknown.

However, it would be logically inconsistent to conclude that the process of biological adaptation to preserve life (evolution) is separate and apart from whatever forces in the universe spawned life in the first place. The two must be intertwined. They must exist concurrently and interact in balance with one another like the pull-and-give of two sides of a rubber band or string.

Evolution is not fully appreciated yet as the universal force of nature it may be. To not include evolution with the other major forces of nature (electromagnetism, strong and weak nuclear forces, and gravity) ignores evolution's impact upon all life. Life itself is a force of nature, interconnected with each of the other major forces in ways we do not yet fathom.

Our theory is that PhiAlpha is an evanescent, emergent genetic trait evolution constructed over time, applying fundamental processes of thermodynamics, physics, and biochemistry to comply with the fundamental conservation laws of energy and information.

The physical universe is a flux of matter and energy that constantly changes states over time. Therefore, the state of existence we humans know as our present physical life is not meant to exist forever in the same form.

By the same laws, life cannot cease to exist either. That would not be logical. Logic tells us that existence continues in some form or state of matter and energy after physical death. The laws of energy and information conservation mandate that information continues to exist—in one form or another—forever.

PhiAlpha is the bridge evolution created to comply with these two laws. Human evolution resulted in individuality as the genetic trait best suited to organize and preserve information under natural law. PhiAlpha is the logical evolutionary adaptation of biological processes needed to form a coherent phased state of energy capable of receiving, prioritizing, storing, and transferring information (memories) in sufficient quantity and quality to isolate and preserve individuality between human physical lifetimes.

Once the physical body dies, incorporeality diffuses the information stored in the brain like steam escaping a boiling pot of water. To conserve information beyond death, evolution needed a method of creating a coherent pulse of information capable of surviving the loss of corporeality until reorganizing itself in a living host.

PhiAlpha must, therefore, be a combination of genes, proteins, and amino acids whose expression need not occur during normal life at all. They may lie dormant during normal life and only become expressed as the body encounters a life-threatening event.

As the body reacts to the threat, a biochemical "trigger" would cause the expression of genes necessary to organize, prioritize, store, and transfer information (memories) from the brain's neural net into a coherent energy state lasting long enough preserve the information as it transits from the dying host and then into a living fetus.

Evolution's capacity to upregulate genes at near death or even postmortem was documented in 2016 by Peter Noble, PhD, a professor at the University of Washington, whose expertise is in the field of gene expression, molecular genetics and DNA sequencing.

In one experiment, Noble studied how zebrafish and mice genes behaved after death. To his surprise, neither sets of genes decayed when their fish and mouse hosts died. Instead, they suddenly began functioning only after death. This presents a rather bizarre mystery. Why certain genetic profiles turn on or are upregulated (including thanatotranscriptome) only after death is unknown:

"Upregulation of genes demonstrates formation of new molecules post-mortem. Sufficient energy within cells of dead animals exits to maintain gene transcription for up to 96 hours after death. An unknown regulatory network appears to be still turning "on" and "off" genes in organismal death..."[19]

Noble's team also found odd the timing of gene upregulation at death. While most of the genes "turning on" at death did so immediately at death, some did not turn "on" for 24 to 48 hours post-mortem:

"These differences in timing and abundances suggest some sort of global network is still operating. ...What makes gene expression of life different from gene expression at death is that post-mortem upregulation of genes offers no obvious benefit to the (now dead) organism. We argue that self-organizing processes driven by thermodynamics are responsible for the post-mortem upregulation of genes."[20]

Noble and his colleagues correctly point out the mystery regarding the upregulation of genes post-mortem without any apparent benefit to the now deceased animal. Evolution is never without reason, however.

Certainly, the ability to turn specific genes "on" at and after death is a necessary component for PhiAlpha to operate. That this capability is found in lower life-forms is evidence of evolution's interest in the process of selective expression of genes at and after death. We see evolution at work, preparing individual processes in lower life-forms to be brought together in higher life-forms.

Evolution is a zero-sum process. Natural selection is all or nothing. If the genetic code of a species maximizes its group fitness for survival (the dinosaurs, for example), it will be able to pass those genes along to its children indefinitely (unforeseen asteroids the exception, of course). If, however, a species lacks group fitness for survival, ultimately there are no children. This species becomes extinct.

Group fitness for survival has been evolution's mantra—until the arrival of humans. Humanity's ascension over the gene-dome as Earth's dominant species caused a fundamental change in evolution's priorities.

Evolution itself is evolving. While continuing to apply its original code of group-fitness, evolution has slowly but gradually built something uniquely human—individuality. Gerald H. Jacobs, PhD, professor of psychology at the University of California at Santa Barbara explains:

"Humanity is in the process of evolving from collective uniformity to increasing variation and diversity. This movement has gained impetus from the growing recognition that the overall strength and sustainability of the collective is proportionate to the value it accords to each individual human being and the active support it lends for full development of each individual's unique, creative potentials..."[21]

Human individuality did not appear out of thin air or overnight. Evolution has been manipulating the trait of individuality for millions, if not billions, of years. True, evolution's methodology may have begun as a shotgun approach toward biological adaptation, but its history reveals its singular purpose to be something quite different.

The major transitions theory of evolution argues that evolution is an ongoing process driven by a small number of major evolutionary transitions. In each transition, a small number of individuals that were previously able to survive and reproduce independently from the group create a new and more complex lifeform. Genes, for example, cooperate to form genomes. Archea and eubacteria formed eukaryotic cells. Cells cooperated to form multicellular organisms.[22]

Evolution has been using individual mutations and group-fitness as an intertwined pair of global trait constructors. One counterbalances the other, but both are essential to the overall process. In 2015, Stuart A. West, PhD, Roberta M. Fisher, PhD, and their research team identified two conditions defining an evolutionary transition in individuality.

Stuart has been professor of evolutionary biology at England's prestigious Oxford University. Fisher is associated with the Department of Ecological Science at Vrije University in Amsterdam, The Netherlands.

First, entities capable of independent replication before the transition can replicate only as part of a larger unit after the transition.

This is called by several terms: mutual dependence, interdependence, contingent irreversibility.

Second, a relative lack of within-group conflict occurs such that the larger group can be thought of as a fitness-maximizing individual (organism). An example of this condition would include a group of cells that make up a multicellular animal as a single organism. When these two conditions are met, West's team hypothesizes, evolution can produce a higher-level individual (organism).[23]

While no species has been individualized more than humans, the trait itself has been in development for untold millions, if not billions, of years. Observation alone confirms that individuality, as a trait, shows increased expression as one moves up the evolutionary ladder toward humanity. The higher and more complex the lifeform the greater the expression of individuality within the species.

As with any new mutation within a species, individuality had to survive as a trait among competing traits, especially the trait of group fitness. In 1998, the noted evolutionary biologist and author of *Darwinian Dynamics*, Richard Michod, PhD, of the University of Arizona, presented an explanation for why individuality prevailed:

"What explains the individuality of organisms? For an organism to be an individual, or evolutionary unit, requires functions at the organism level that protect it from conflict within...natural selection at any level requires heritable variations in fitness at that level. To emerge as an evolutionary individual, cell groups must acquire the means to guarantee fitness heritability at the group organism level."[24]

For evolution to protect its survivability, individuality must have appeared as a very important trait necessary for survival of complex species. To support this theory, Michod identified germline modifiers and self-policing modifiers as the first uniquely organismal functions that evolve during the transition from unicellular to multicellular life. These modifiers protect the evolved (mutated) cell from being attacked from within the pro-unicellular enzymes.[25]

We humans are billions of years ahead of simple multicellular organisms, however. How evolution has modified us over those eons of time remains unclear.

However, Cory McLean and a team of Stanford University researchers have been attempting to identify molecular events particularly likely to produce significant changes in human genetic structure—such as deletion of DNA sequences that were otherwise highly conserved between chimpanzees and other mammals. They were able to identify more than 500 such deletions in humans, which are almost exclusively from what are called "no coding zones." He concludes:

"Humans differ from other animals in many respects of anatomy, physiology and behavior; however, the genotypic basis of most human-specific traits remains unknown."[26]

Evolution and Individuality

We define the term "individuality" to mean the identity matrix created when we subjectively prioritize those memories upon which our self-image depends.

This identity matrix changes over time as we acquire new experiences and emotional responses to those experiences. It would take days to go through all our memories to isolate only those upon which we identify as necessary to describe who we are. PhiAlpha collects, reviews, and prioritizes an individual's memories almost instantly.

(1) Individuality is more important than most other genetic traits such as hair color, eye color, facial features, bipedality, opposable thumbs, and the folds and ridges of the human brain's cortex for greater brain area and greater complexity of thought despite the physical limitations of the human skull. Indeed, individuality is equal with natural selection and group fitness as fundamental algorithms of evolution.

(2) Evolution has been refining ever more complex forms of individuality from the time multicellular organisms emerged from the unicellular. We cannot expect evolution to stop now.

(3) Individuality has proven its survivability over time and must be recognized as a dominant trait. Recorded history confirms individuality's irreversibility as such a trait. Despite repeated efforts over history, no tyrant nor dictator has yet been able to crush the concept of individual freedom from human consciousness. Individual freedom of expression is an irresistible urge within the consciousness of all humans—an urge evolution has gone to great effort over eons to create and to protect.

Evolution has one major problem with individuality, however. The more complex the brain becomes, the more difficult it becomes to pass along individuality through one's genes. Sexual reproduction requires two participants and is, therefore, a group-oriented process. Eventually, individuality no longer remains a group process but begins to be isolated within individuals. The more complex the brain, the more specific the experiential memories become.

The concept of sexual reproduction is a group-oriented process requiring more than one individual to carry on the genetic code of the species. This process creates a need for individuals to interact as part of a social order or group to find mates. Individuality is revolutionary in the sense that it no longer requires group sanction or group interaction to thrive or to survive. Individuality is anti-group evolutionarily as well as socially.

At some point, the development of human individuality conflicted with two of the laws we mentioned earlier: conservation of energy and information. This likely occurred when brain complexity permitted sufficient memory storage to make these memories more valuable to the survivability of the human species than traits passed along genetically from one generation to the next (eye color, physical features, etc.).

How does one pass along a memory from one generation to the next?

Cloning does not work. We can clone the genes that make up one's body and even brain structure, but we cannot clone one's individual life experiences. Even if we could do that, we could not effectively clone an individual personality.

Human individuality cannot be cloned because each person prioritizes memories differently by subjectively chosen criteria. We are the product of the experiences we subjectively choose to motivate our emotions and behavior. One cannot duplicate another's choices by cloning.

Nor can we "download" our memories to the next generation genetically. Personality cannot be transferred to progeny through procreation. Parenting and socialization are the only ways that the identities of the individual and of society can be transferred from one generation to another.

Therefore, if individuality is a necessary trait evolution needs to preserve to comply with the laws of conservation of energy and entropy how does evolution construct such a process as PhiAlpha?

Evolution and the Human Brain

The answer lies in the evolution of the human brain. Evolution has made humans special. We are special because of our brain. Recent research bears this out. The human brain size tripled 1.5 million years ago, with the cerebral cortex expanding at a much faster rate than other areas of the brain.[27]

The human brain is no longer organized the same as the brains of other mammals and primates. This reorganization of central components of the human brain is as important to the brain's development as was its increase in size.[28] It has been suggested that this reorganization occurred due to the human brain's need to manage increasingly complex social and interpersonal interactions and relationships.[29]

In addition, Todd Preuss, of the Yerkes Primate Research Center at Emery University, found more than 100 separate genes differentiated human and chimpanzee brains. He discovered that human brains possess remarkably different spatial patterns of genetic expression compared to both chimp and macaque brains.

Preuss admits he does not yet understand what these differences mean in terms of functional organization, though he does suspect that many of these changes involve upregulation of genes in areas controlling human metabolism, synaptic organization, and synaptic function. These genes are creating a more dynamic brain. As Preuss put it, "The human brain seems to be **running hot** in all sorts of ways [emphasis added]."[30]

If the human brain is indeed "running hot," we do not yet know where it is running to—or why. We do not know evolution's ultimate purpose. Some suggest evolution has no purpose. They are wrong. No biological process exists without purpose. Natural selection makes any other outcome logically impossible. Even if the only purpose of evolution were to be species survival, then that is a purpose.

We do not know evolution's ultimate purpose, but we can speculate. Evolution never operates without a specific plan based upon improved functioning it deems necessary for increased survivability. Research suggests that evolution is directing humanity toward unlimited cognitive and communicative capabilities with the intention of enhancing the expression of individuality within each of us.

Evidence of this is documented in the research of Chet Sherwood, PhD, and his team. Sherwood has been professor of anthropology at The George Washington University in Washington, D.C. After obtaining his doctorate from Columbia University, Sherwood has researched brain evolution in mammals and is affiliated with GWU's Mind-Brain Institute for Neuroscience.[31]

Sherwood identified genetic differentiations between the human brain and those of our primate and mammalian ancestors. The most significant difference between humans and lower species is that of language. He and his team concluded that language is "a unique behavioral adaptation in humans. ...[T]he unique brain growth trajectory of modern humans has made a significant contribution to our species' cognitive and linguistic abilities."[32]

From this research, we can conclude evolution appears to have chosen individuality as a dominant trait in and of itself. Moreover, evolution has chosen the human brain as the most efficient means in which to preserve and refine individuality further. We can express this relationship as a formula:

$$I = (C_1 \ L_1 \ M_1 \ / \ t)$$

Where the letter I stands for the completed trait of individuality; C for cognition; L for language; M for memory and t for time. The subscript 1 below C, L, and M symbolizes a sufficient complexity of these variables at a level sufficient to achieve individuality. Cognition, language, and memory are all evolutionary components of intelligence, and their interaction over time creates individuality.

Why is individuality so important to evolution's plan for humanity? We suggest that the reason is found in the path evolution has taken from primordial times. Evolution is not sentient in the way humans are, with subjective, experiential involvement that mandates varying degrees of bias. Evolution has no subjective identity or intention.

While evolution may use survivability to separate the wheat from the chaff, so to speak, it seems to being using hidden algorithms to change us into something far different from the rest of Earth's species.

One such hidden program, as it may be called, is focused upon ever-increasing awareness or sentience. Natural selection and survivability are just means to an end. They are routers, so to speak, guiding evolution away from species that it would be wasting its time upon and toward a species most suitable to use expanded awareness as a tool and to survive and thrive with this tool.

Individuality is the ultimate application of increased awareness. Evolution has chosen humanity as the only species possessing a brain of sufficient complexity and size to express itself in art, music, song, poetry, literature, religious rituals, sport, and complex forms of interpersonal communication. These traits require unique expressions of individuality.

The greater our awareness of ourselves and our environment, the greater ability we possess to express our individuality. Every emotion that evolution has crafted for humans only serves to enhance and enrich the individual experience of life. Our universe is a symphony of mathematically eloquent logical sequences of ever-increasing complexity.

It is, therefore, logical that evolution would not permit human individuality to be lost. If evolution spent billions of years creating ever more complex means of individual expression, it is logical that evolution would not accept losing its creation to the transience of human lifetimes. There must be a means of isolating that individuality and then transferring it from the body of the dying to that of the living.

How can that happen?

5. INDIVIDUALITY: CHILD OF OUR MEMORIES

To answer this question, we begin with the building blocks of individuality—memories. Memories are experiences created, stored and transmitted through neurons and synapses in our brain. Memories are what the human brain uses to craft the uniqueness of our individual personalities over time. Individuality requires both working memory (day-to-day memory) as well as long-term memory (those memories stored over years or even a lifetime). Working memory allows us to function normally through our daily routines, carry on meaningful relationships, and make rational decisions as we live our day-to-day lives.

Nevertheless, we suspect the memories PhiAlpha prioritizes to preserve individuality are long-term memories. Long-term memories are stored in the cortex, in what is called the gray matter. It is this area of the cortex that appears in photos and diagrams as though folded over upon itself.

Evolution caused the human brain's cortex to fold in upon itself to create more area for greater computational power within the confines of the human skull.

The gray matter is where all the neurons and synapses are located. The folds create ridges called gyri (gyrus, singular) and valleys called sulci (sulcus, singular). Underneath the gray matter is a layer of white matter, which is composed entirely of fibrous axons that act as wires connecting the neurons of one fold to those in another fold.

In healthy humans, these axons are myelinated (covered in an insulating sheath allowing the signal to travel easier and faster).[33] However, at least one study suggests long-term memories are stored in very specific brain cells located in only one area of the human brain—the dentate gyri of the hippocampus.[34]

To fully appreciate the interaction of PhiAlpha and individuality, it is necessary for a brief review of how long-term memories are created and stored. Processing loads within the human brain require an astronomical number of neurons, synapses, and glial cells. The human brain contains over 200 billion neurons with up to 10,000 types of neurons. Each neuron is connected to between 5,000 and

200,000 other neurons. It has been suggested that there are more neural connections in the human brain than there are stars in the universe.[35]

The neurons interact with synapses, axons, and glial cells as signals are transmitted from the brain to and from parts of the body. Synapses carry information by electron and/or biochemical ion transfer between neurons. Neurons and synapses are mostly located in the gray matter of the cortex (the part of the gray matter that is folded creating gyri and sulci). Axons are in the white matter part of the brain lying beneath the gray matter.

We can regard neurons as batteries of stored electrochemical energy. Synapses are like switches, turning on and off as needed to transmit the electrochemical energy into signals. The axons are the wires over which the signals travel. It reminds one of the structure of our modern microprocessors, doesn't it?

It works in the same way. All cells in the body function through electric polarization of molecules. Ion transfer across cell membranes affects voltage levels, which result in either polarization or depolarization of the cell. Memories are contained and moderated in the cells of neurons and synapses. Each neuron may be an individual event memory.

Memory capacity of each neuron is determined by the strength and size of its attached synapse. Synapse size is dependent upon how often each synapse is "fired" by an electrical or biochemical ion signal called an "action potential." Action potentials (more commonly known as nerve impulses) are created by both the voluntary (conscious thought) nervous system or the involuntary (no thought involved) nervous system. Action potentials "fire" when sufficient energy is created within the neuron to send a command or when we relive an experience or consciously think about something. Therefore, synapses increase and decrease in size depending upon how many times the brain uses them.[36]

Researchers estimate every synapse in the brain increases and decreases in size every two to 20 minutes. Total memory storage of the human brain depends upon both the number of neurons and their synapses and the number of distinguishable synaptic strengths. Each synapse is estimated to possess the equivalent of 4.7 bits of computer memory.[37] Multiply 4.7 bits of data per synapse by the number of possible neural-synaptic connections, and the total amount of memory storage space in the human brain is staggering.

All neurons send and receive messages electrochemically by metabolically powered ion pumps that combine with ion channels inside the cell's membrane. These channels generate differentiated electrical charges (interchanging between positive and negative charges of varying strengths) between the interior and exterior of the neurons. Theses electrical charges are created by ions (sodium, potassium, chloride, and calcium atoms, which possess differing levels of electrical charge).[38]

The change between positive and negative ion charges creates a biological alternating current

similar to that found in the common motor. This process creates the energy needed within the cell to pump or move the neural signals from one part of the brain to another.

When an ion crosses the boundary of a neural membrane, the voltage it releases fires an action potential. The voltage required for the firing of an action potential is about -55mV (millivolt). The energy pulse released by the firing of the action potential travels along the cell's axon, activating synapses with other cells (including other neurons).[39]

The electrochemical process that allows the neuron (and all other cells) to transfer electrical signals is called the sodium-potassium pump. The push-pull of osmosis into and out of the cell by this "pump" causes differences in the electrical charge between the interior and exterior of the cell membrane. This push-and-pull effect forces sodium ions within the cell to be repelled outward while potassium ions are propelled inward.

Until recently, it had been assumed that the only function of the sodium-potassium pump was to monitor and control local voltage levels within and around cell membranes. However, the sodium-potassium pump is suspected of being much more than that. Some researchers believe the pump may control cerebellar neurons and may also serve as a computational processor or computer in both the cerebellum and cortex.[40]

Memory recall is a complex electrochemical process. It is this process that creates what is called the brain's bioelectric field. This bioelectric field is created as follows:

When memories are recalled from neurons in the gray matter, their signals travel downward from the gyri of the gray matter into the white matter. Once in the white matter of the brain, axon fibers route the signal over and then upward from the white matter back into the gray matter. Now back in the gray matter part of the brain, the signal is transferred to the appropriate neurons connected to those memories that are stored in other folds within the cortex.

This routing of electrochemical neural signals downward into the white matter and then back upward into another area of gray matter takes the electric field created by the transfer of electrochemical ions in the cortex and spreads it out into the cerebrospinal fluid pulsing over and through the cortex.

Cerebrospinal fluid is an electrically conductive medium, which means it transfers an electric current easily. We believe the down-across-up motion of neural electrical signals through the wavelike folds of the cortex generates an electromagnetic wave through which neural information flows like a current.

This electromagnetic wave interacts with the electrically conductive medium of the cerebrospinal fluid (CSF) to generate and sustain an electromagnetic field capable of supporting a state of phased energy within the brain and spinal cord we term the neural network (neural net). We suspect this neural net serves as an antenna transmitting and receiving biophotons containing individuality in the form of memories stored within the biophotons as coherent bits of information.

How is this accomplished?

We have examined evolution's emphasis upon brain capacity and its relationship with cognition, language, and ultimately expanded awareness leading toward ever-increasing complexity of individuality. We have explored the relationship between individuality and memory, and how memories are constructed. We have demonstrated why individuality cannot be cloned or transferred genetically from one generation to another like eye color or asthma.

If PhiAlpha exists, it does so not as a mystical force independent of physical laws of nature or the universe but as a child of universal laws and processes dependent upon them. PhiAlpha would obey the laws of natural selection and fitness for heritability and survival just as any other process evolution creates. Because PhiAlpha would arise from and out of our own physical bodies, it would obey the same laws of thermodynamics and quantum mechanics as any other physical state would and does.

We cannot expect PhiAlpha to "shout out" its existence at us with one or several specific clues to its existence. PhiAlpha is the forest hiding amid the trees, so to speak.

To find PhiAlpha, we must search for its fingerprints upon the common, ordinary physiological processes we take for granted every day. We need to know what to look for. To do this, we must hypothesize what biological functions PhiAlpha would need to accomplish its task of preserving individuality as memory information. Then we must look within the human body to see if there is any evidence of such processes. If those processes exist, we can deduce PhiAlpha's existence, even if we are presently unable to observe its behavior.

What is the purpose of PhiAlpha?

We understand that evolution employs group fitness and natural selection with the individuality of single mutations to create a counterbalance of spiraling complexity in organisms. We also understand that evolution has selected *Homo sapiens* as a species to be separated from the herd of other species for special attention. This special attention includes maximizing complexity of cognition, language, and memory—creating ever more complex expressions of individuality.

Individuality has not only become the most important trait evolution has created for humans, human history has shown that individuality has become an end unto itself. Therefore, the purpose of PhiAlpha must be the continuation of coherent individuality stored within neural-encoded memories.

PhiAlpha is more easily understood by differentiating its component processes. PhiAlpha components include: (1) **PhiAlpha Activation**; (2) **Memory Organization**; (3) **Memory Prioritization**; (4) **Memory Encoding**; (5) **Memory Emission** and (6) **Memory Reception**.

We will examine each in order.

6. PHIALPHA ACTIVATION

PhiAlpha cannot be "on" all the time, otherwise its system requirements would dominate other physiological systems prohibiting normal life function. PhiAlpha must, therefore, lie dormant during normal life events and require a specific genetic-upregulating stimulus or event for expression.

More simply put, PhiAlpha cannot become active every time we catch a cold or get the flu. If that were the case, our bodies would not be able to function. Even abnormal events such as a broken bone or disease will not initiate PhiAlpha. While serious, each of these conditions are survivable with proper medical care.

Rather, PhiAlpha activation must require an extraordinary life-threatening event. Further, PhiAlpha cannot be a simple one-step or even a two-step process. If this were the case, our minor illnesses and routine accidents would cause PhiAlpha to rev up too easily and too soon. The added physical stress of PhiAlpha's start-up would only exacerbate otherwise routine illnesses and injuries.

To guard against premature PhiAlpha start-up, evolution would need to create a specific set of conditions necessary to trigger PhiAlpha as well as a biological mechanism to control its performance. Therefore, PhiAlpha would need to distinguish between a life-threatening but reversible crisis and the beginning of the death process itself.

Evolution has created the perfect mechanism to serve as a PhiAlpha trigger and monitor—the brain's limbic system:

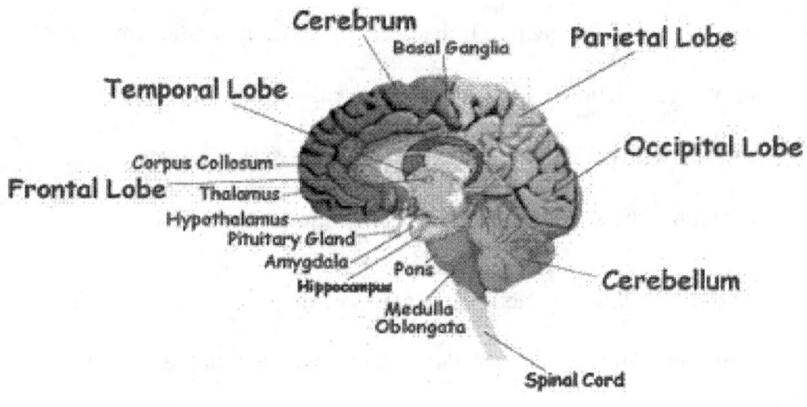

Illustration No. 1

Illustration No. 1 reveals the limbic system (hippocampus, thalamus, amygdala, hypothalamus, basal ganglia, and cingulate gyru) in relation to other organs of the human brain. The limbic system is located in the most secure area of the brain—at the brain core, near the brain stem, itself.[41]

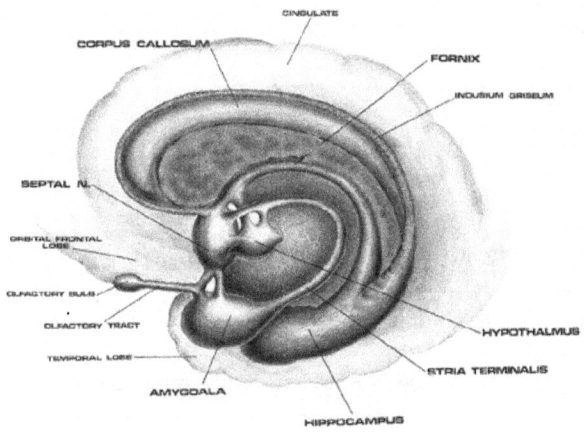

Illustration No. 2

Illustration No. 2 provides a close-up view of the separate organs of the limbic system and their relative positions and sizes. Note the

95

complex and intricate shapes which make up the various organs of the limbic system. Overall, the limbic system appears crescent shaped, similar to that of the human ear. The illustration highlights how the limbic system is composed of an intricate arrangement of many brain organs. While the size of these organs may be small when compared to the rest of the brain, their functions and interactions are extremely complex.

The limbic system is critical to our theoretical model because it is the command and control center for all PhiAlpha processes in the dying body. We will shortly see that the limbic system's control over essential body functions, consciousness, emotions, and memories will have a direct impact upon PhiAlpha's critical functions. But before we can adequately explore the connection between the limbic system and PhiAlpha, we must first learn some basic information about each of the organs making up the limbic system.

The major organs of the limbic system are:

1. **Amygdala**

2. **Cingulate gyrus**

3. **Corpus callosum**

4. **Fornix**

5. **Hippocampus**

6. **Hypothalamus**

7. **Indusium griseum**

8. **Septal nuclei**

9. **Stria terminalis**

We will learn more about the various organs of the limbic systems as we proceed, but for now we need some basic information on each:

Both the **amygdala** and the **hippocampus** are bulbous in shape, though the hippocampus is at least twice the amygdala's size. They are both located at the base of the limbic system, immediately adjacent to the temporal lobe. The hippocampus processes new memories for long-term storage and regulates the mind's sense of present awareness.[42]

Protruding out of the amygdala is the **stria terminalis**, which carries nerve impulses to and from the septal nuclei, hypothalamic, and thalamic areas of the brain. The stria terminalis a ribbon-shaped band of nerve fibers connecting to the septal nuclei, amygdala, hypothalamus, and thalamus. It is believed to serve as a means of implanting fear and anxiety impulses upon memories.[43]

The **cingulate gyrus** sits just above the corpus callosum. Like many of the other organs of the limbic system, the cingulate gyrus is crescent-shaped and surrounds the other organs of the limbic system. It looks almost like a helmet as it sits atop the other organs of the limbic system.[44]

The **corpus callosum** is a thick crescent-shaped band of nerves connecting the left and right hemispheres of the brain. The corpus callosum is the largest nerve bundle in the brain, containing between 200 million to 250 million axons. The corpus callosum possesses the largest amount of white matter in the brain and is located between the cingulate gyrus and the fornix.[45]

The **fornix** is a crescent-shaped nerve bundle that serves as the main output channel for the hippocampus. It also carries nerve fibers to the hippocampus from the forebrain. Like the rest of the organs of the limbic system, the fornix is known to play an important role in how specific emotions are bonded upon our memories. Like the rest of the organs of the limbic system, not a lot is known about what other functions the fornix may have.[46]

The human **hypothalamus** is located at the base of the brain, near the pituitary gland. About the size of an almond, the hypothalamus is positioned almost in the exact center of the skull.

It is believed to control or mitigate a large number of brain functions, including pituitary gland function, hormonal release, circadian rhythm, body temperature, and behavior. The hypothalamus is highly reactive to light, olfactory stimuli, steroids, neurally transmitted signals from the heart and other major body organs, blood-born stimuli, infectious agents, microorganisms, and stress.[47]

The **indusium griseum** is a thin layer of gray brain matter covering the upper surface of the corpus callosum. It contains four longitudinal bundles (two on each side). Again, not much is known of its function. Its location suggests it plays an important role in the limbic system's emotional-memory organization.[48]

The **septal nuclei** are a set of structures made up of both a lateral and medial component. Bulbous in shape, they are located below the corpus callosum and are connected to the olfactory bulb, hippocampus, amygdala, hypothalamus, midbrain, cingulate gyrus, and thalamus. They are believed to play a role in pleasure responses, but the literature indicates an almost complete lack of understanding of its functioning.[49]

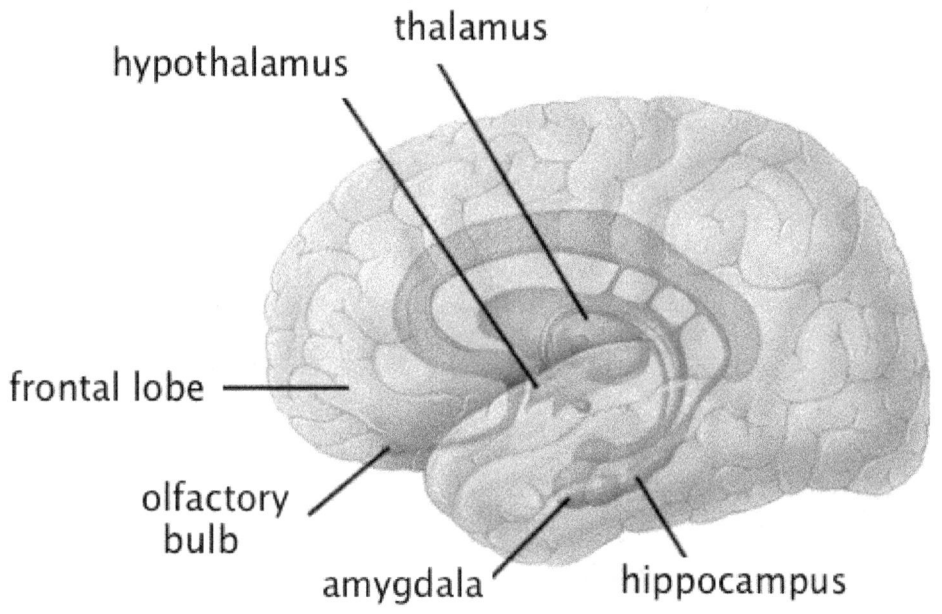

Illustration No.3

Illustration No. 3 shows the position of the thalamus in relation to other major organs of the limbic system as well as its location within the brain itself. Interestingly, this diagram depicts the limbic system as a serpent, with the amygdala and hippocampus as its head and the rest of the limbic system as its coiled body. It does not take much imagination to notice the similarity between this serpent-like shape, its connection to our emotions and our emotional behavior, and the biblical story of Adam and Eve. Is it possible this story is but an allegorical fable meant to explain to an unsophisticated agrarian culture the effect of their emotions upon their daily lives?

The **thalamus** is a knob-shaped structure of gray brain matter with twin lobes located in the center of the brain. While not considered by current medical science to be a part of the limbic system, its extensive array of nerve connections to the hippocampus suggests otherwise. It also has connections with the cerebral cortex and is believed to play a significant role in consciousness and states of sleep and wakefulness. The thalamus acts as a relay station—sorting, processing, and directing signals traveling between the spinal cord and the rest of the body.[50]

The **spinalthalamic tract** is a nerve-fiber pathway allowing signals to be transmitted between the thalamus and the spinal cord. The thalamus is believed to use the spinalthalamic tract to receive nerve impulses from the spinal cord. This connection can be better understood by examining Illustration No. 1, which shows how the spinal cord is directly connected to the midbrain and the limbic system. It is clear from Illustration No. 1 that any information going between the cerebral cortex and spinal cord must first pass through the limbic system. Later, in the chapter on memory absorption, the relationship between this limbic system and the spinal cord will provide important clues to how memories from our past lives are absorbed by our new bodies.[51]

It must be emphasized that as much as modern science has advanced in the last 50 to 100 years, it has yet to unlock the many mysteries within in the brain and the limbic system. While it is understood that the limbic system plays a critical role in how our emotions are imprinted upon our memories of specific events, the scope and specific other functions of the limbic system—and the individual organs themselves—remains largely unknown. It is a simple fact that with every single piece of knowledge we gain about the brain, we learn there is a great deal more about what the brain does that we do not yet understand.

For example, while researchers have known for some time that the brain possesses both electrical and magnetic fields, they have only just begun to study how to validate and calibrate electrical-current flow in computer models with the results taken from living human test subjects.

We have discussed how cerebralspinal fluid (CSF) is an excellent source of both electric field and magnetic field activity in the human body. The reason for this is that CSF contains many elements that, when dissolved in CSF become electrolytes such as sodium, potassium, magnesium, chloride, phosphate and bicarbonate. CSF also carries salt water and trace elements, including copper, zinc, aluminum and iron—all of which serve as excellent electrical conductors.[52]

We call the electromagnetic field generated by CSF flow the "neural net." This neural net is a biomachine created by the pulsing action of the CSF flow creating what is essentially an organic computer. As with any man-made computer, our brain runs on various programming algorithms to perform its multitude of functions.

One such function is PhiAlpha. The brain's PhiAlpha program is activated upon a significant crisis threatening our lives. Regardless of the proximate cause of our death, the initial point at which our body becomes aware of the possibility of death is what is called the "stress response," and that begins in the amygdala.[53] An almond-shaped mass of interconnected neurons located deep in the brain's medial temporal lobe, the amygdala modulates our reactions to events that we define as critical to survival.

The amygdala controls our reactions to all sorts of major events in our lives, from signaling the presence of food to warning of predators and rivals, the selection of mates to the instinctive responses to the threat of danger to children and similar crisis situations.[54] Any event the senses perceive as a credible threat to survival triggers an immediate response from the amygdala.

The amygdala's neurons may serve as the memory store necessary to confirm or reject the sensory stimuli alleging the threat. In other words, the neural memories contained within the amygdala may serve to either verify the stimuli as a valid threat or not, depending upon how the new threat stimulus matches with memories of past experiences. Certainly, to function properly as the brain's early-warning center for life-threatening events, memories contained in the amygdala would have to be very long-term memories indeed.

Upon validating a true crisis, the amygdala sends a signal to the hypothalamus. The hypothalamus is also almond shaped and is located near the amygdala in the brain's most protected core area: next to the pituitary gland, just above the brain stem.

The hypothalamus acts as the central processing unit of the body's autonomic (involuntary) nervous system. It maintains the body's internal chemical and electrical balance, and controls such functions as heart rate, blood pressure, body temperature, fluid and electrolyte balance, appetite, body weight, glandular secretions of the stomach and intestines, and sleep cycles via the release of hormones.[55]

There are few, if any, threats to the life of the body that are not detected by the hypothalamus. All these functions occur without our conscious awareness. Both amygdala and hypothalamus work together to instantly respond to crises before our brain's vision center is able to cognitively identify the event as a threat.

The following flowchart illustrates the physiological event cascade that results from the detection of any possible threat by the amygdala and hypothalamus:

1. Signal from amygdala to hypothalamus.

2. Signal from hypothalamus to involuntary nervous system (NS).

3. Signal from involuntary NS to sympathetic NS.

4. Signal from sympathetic NS to autonomic N.S. and the adrenal gland.

5. Adrenal gland creates and releases adrenaline (epinephrine).

6. Adrenaline release causes physiological changes (increased heart rate and blood pressure, and senses become significantly more focused.

7. Adrenaline is exhausted causing hypothalamus to activate secondary hypothalamus, and the pituitary and adrenal glands' (HPA axis) threat response.

8. HPA axis signals release of corticotropic hormone (CRH) into the pituitary gland, which then triggers release of adrenocorticotropic hormone (ACTH) that travels to the adrenal glands and initiates release of cortisol.[56]

Cortisol is the body's last line of defense against a life-threatening event. A complex hormone released only after initial release of adrenaline has become exhausted, cortisol is the body's final biochemical stimulus to enhance performance to counter any perceived life-threatening event. Cortisol release initiates a series of physiological changes to help us survive. Some changes are easily understood. However, the purpose of several other changes created by cortisol remain mysterious.

For example, cortisol upregulates (turns on) three genes: interleukin-4; interleukin-10; and interleukin-13. These genes help the body reduce the severity of injury by increasing anti-inflammation and antibody production. As a diuretic, cortisol increases water excretion through perspiration and renal plasma flow from the kidneys.

These are just a few of the many effects of cortisol. Remember, cortisol is released only after the body has expended its entire supply of adrenaline without reducing the threat level.[57]

However, cortisol also increases sodium retention by causing the sodium-potassium ion pump to go into overdrive. When death seems imminent, cortisol forces increased retention of sodium ions at the same time it is trying to increase water excretion. Sodium atoms retain water. These processes are in opposition to each other. They conflict.

What is happening? What is so important about the cellular sodium-potassium ion pump that it forces cortisol into causing two opposing physiological processes during a time when one would logically expect all physical processes to work together, not against each other?[58]

Before we can answer this question, we must first discuss cortisol's other effects.

Cortisol has two other effects upon the body in crisis that are significant to our understanding of PhiAlpha. It stimulates the release of copper enzymes including lysyl oxidase and elastin. While copper enzymes maintain and repair connective tissue that may be damaged during the crisis, copper is also a perfect electrical conductor. By release of copper enzymes throughout the body, cortisol may be preparing the brain's neural net for the electrochemical reactions necessary to activate PhiAlpha.

This would also explain the sudden and seemingly contradictory retention of sodium ions in neural cells. Sodium ions are also an excellent electrical conductor.[59] PhiAlpha relies upon the body's electrical and electromagnetic fields to encode memories upon phased biophotons as they are released from cells within the brain.

Lastly, during extreme physical crisis, cortisol causes the brain to create what have become known as "flashbulb memories" of the crisis event.[60] We suspect these crisis memories are given this flashbulb physiological emphasis to hasten their routing to the amygdala and encoding into PhiAlpha if the crisis becomes fatal or preserved for future reference in the amygdala's long-term memory storage.

Illustration No. 4

Illustration No. 4 presents the moment of impact of a motor vehicle accident. Images such as this become part of the flashbulb album of memories the limbic system measures for the emotional weight we give them.

We conclude that it is this release of cortisol that triggers activation of PhiAlpha in the brain. Being at the heart of the body's central processing unit, the limbic system is perfectly designed to monitor the crisis and modulate the expression of PhiAlpha accordingly. In the event the body is mortally wounded, the limbic system has already set in motion those functions necessary for PhiAlpha to take over. If the crisis is averted, however, the limbic system discontinues cortisol release. PhiAlpha is shut down without the individual ever being aware it existed.

7. MEMORY ORGANIZATION

PhiAlpha is initiated once the limbic system becomes aware that death may be imminent. But PhiAlpha is a multistage process. The initial stage of PhiAlpha is memory organization.

Survivors of near-death experiences often describe a period during which they observe their lives flashing before their mind's eye. We suspect this intense flood of visualized memories—some of which are from childhood and, perhaps, even from past lives—to be direct physical evidence of this stage of PhiAlpha. We also suspect there to be a connection between the limbic system's flashbulb memories and the experience of having "life flash in front of your eyes," as has been anecdotally reported.

Biologically, PhiAlpha must perform its functions within a narrow time window. Once the heart stops, oxygen is no longer pumped into the brain through the bloodstream. Brain damage begins within three to five minutes after cardiac arrest. Brain death occurs within minutes thereafter (allowing for body temperature, causation, and extent of any physical trauma, age, chronic disease and other variables).[61]

As a metabolic process, PhiAlpha requires energy to function. This energy must come from the brain cells involved. Though even a corpse possesses a continuing ability to upregulate genes, we believe there is a bright-line, a time limit or point of no return after which PhiAlpha would no longer function due to loss of oxygen in the brain's blood supply.

Upon cardiac arrest, this energy supply is limited to whatever enzymes and other energy forms the cells can obtain internally. Once this energy supply is exhausted, the brain can no longer support the metabolic needs of downloading, prioritizing, encoding, and transmitting of memories required by PhiAlpha.

Before we discuss memory organization further, it is necessary to review how memories are made, stored, and retrieved. The issue of memory construction is relevant, because whatever the process of memory storage happens to be, we believe PhiAlpha reverses this process to organize those memories its processes identify as those necessary for individuality preservation. Science still does not understand all the basic memory processes.[62]

What is known is that there are three types of memories: ultra-short-term sensory stimuli memory; working memory for decisions and day-to-day living; and long-term memory. Events chosen for memory encoding are selected by cooperative processes of the thalamus and amygdala as they occur. At the same time the thalamus regulates neural firing of action potentials to encode the event as a memory upon a neuron, the amygdala creates an emotional imprint upon the neuron prioritizing its emotional significance.[63]

The greater the emotional significance to the memory, the higher the encoded intensity within the neuron. We suspect the amygdala controls encoding of emotional intensity by varying the strength and duration of the synaptic firing.

Memories are then stored in the appropriate sensory lobes of the cerebral cortex and regulated by the hippocampus to create a fluid memory sequence when recall of an event is required.[64] This is analogous to a motion picture. These edits are so well hidden, the observer never knows they occurred. Because emotions are unique to the individual, memories of events can vary greatly even among those who witness the same event. This is one reason testimony of even honest witnesses at trial can be very inconsistent and even contradictory. The hippocampus determines which memories are to be selected for storage as long-term memories.

How this is accomplished is not yet known.[65] The exact mechanism of memory construction is not thoroughly understood. However, it is believed to be related to electrochemical energy transfer inside the neuron itself. It is believed that a process called "long-term potentiation" allows the synapses to fire action potentials in sufficient number and intensity appropriate to the significance given the event to be placed in memory. This long-term potentiation is in the form of discharges of high-frequency electrochemical energy by the synapse or synapses involved.[66]

Additionally, one of the lesser known functions of the hippocampus appears to be the intensification of the brain's electromagnetic field at death. Experimenting with new technologies for detecting hard-to-find trace metals using laser ablation combined with mass spectrometry, Researchers J.S. Becker and Myroslav Zoriy of Forschungszentrum-Julich (Julich Research Centre), a leading German interdisciplinary research center in Julich, Germany, in 2005 found copper in higher overall concentrations in the hippocampus, including discrete copper concentrations in the dentate gyrus, and in higher concentrations in the layers of the cornu ammonis (located between the dentate gyrus and subiculum).[67]

The presence of copper within the limbic systems itself is evidence of direct involvement in the electrical conductivity needed to create an electromagnetic field.[68] But an electromagnetic field requires a continuous current of electricity to exist. Does the brain possess such a constant flow of electrons around the cortex that would link up all the memories in all the lobes into a singular, individual state of coherent energy?

The answer lies within the cerebral spinal fluid itself.

Cerebrospinal fluid (CSF) is a colorless, slightly alkaline (pH 7.28-7.32) liquid consisting of water and electrically conductive and ionizing minerals, including sodium, potassium, chloride, calcium, and magnesium.[69] The CSF occupies the subarachnoid space and the ventricular system around and inside the brain and spinal column.

This subarachnoid space contains the layers of delicate meninges tissues within which the CSF flows.

Illustration No. 5 Show how the CSF surrounds the brain like a

Illustration No. 5

river of electrolytic fluid. CSF is always in motion. It is this constant motion of electrolytic fluid that creates the electrical and magnetic field enveloping the brain. MRI video confirms the Klorica and Oreskovic (2014) hypothesis that CSF flow is not unidirectional circulation but the result of cardiac-cycle-dependent bidirectional systolic-diastolic pulse washing forward and backward within the cranium.[70]

Darko Oreskivic and Marijan Klarica are both researchers at Ruder Boskovic Institute (RBI) at Zagreb, Croatia. RBI is the largest Croatian research center in natural sciences and technology.

Elementary particles (electrons and photons) are emitted from the transfer of ionized sodium and potassium atoms across neural cell membranes as the CSF pulsates back and forth over the cortex. This process creates a neural net of coherent information existing in a quantum mechanical state.

Evidence of this neural net was found by researcher S. B. Bauman, of the Department of Neurological Surgery at the University of Pittsburgh Medical Center. In only the second reported study of CSF conductivity, Bauman and his team found that in the seven patients sampled, CSF was found to be 23 percent higher at body temperature than at room temperature as ionic motion and diffusion increased. Further, a 3-D finite-element model of the head using more than 450,000 elements showed that scalp electric and magnetic fields are both highly sensitive to CSF conductivity.[71]

This CSF wash-cycle action across the folds of the cortex creates an electromagnetic field producing quantum mechanical effects throughout the brain—and the mind. It is proposed that this electromagnetic field creates a quantum coherence that unifies all memories together in a singularity of self-awareness—consciousness.

How is PhiAlpha initiated?

Once the heart stops, the limbic system initiates PhiAlpha as follows:

The hippocampus receives impulses from the neural net for those memories controlled by the largest and strongest synapses. These memories are released from their neurons in the form of ions released from ion channels. Ion channels are proteins that make it possible for ions within cell membranes (including neuron and synapse cells) to move across the membrane and out of the cell.[72]

Memories are transmitted instantly to the neural net created by the electromagnetic field of the CSF. The neural net then instantly transfers these memory signals to the limbic system for the next phase of PhiAlpha to begin.

It is important not to confuse this initial "download" of memory-encoded ions as the final PhiAlpha memory product emitted at death. This download is a massive release of memories that need prioritization by the limbic system's emotional control center.

8. MEMORY PRIORITIZATION

We begin this stage of PhiAlpha with several assumptions. First, PhiAlpha cannot transmit the brain's entire memory core at death. This is conjecture, but reasoned conjecture, given the limited time in which PhiAlpha must complete the processes necessary to transmit individuality by release of memory-encoded biophotons prior to brain death of the host and the amount of memory storage the brain possesses.

As discussed earlier, the brain has as many or more neural-synaptic connections as there are stars in the universe. Therefore, there must be a means by which the limbic system evaluates memories to determine which ones are to be transmitted and which ones are not.

Secondly, we theorize that PhiAlpha's transmission capability will not be the same for every individual. Physical defects of age, trauma, disease or other impairment would likely inhibit PhiAlpha from functioning as intended in each death. Further, the energy level of the biophoton burst we suspect is used to release the memories from the cranium at death would also be limited by the same variables. Coherence of the biophotons would be affected as well.

The analogy that comes to mind is a sinking ship full of passengers.

Who do we save?

Who do we leave behind?

The dying body is not a sinking ship, but the problem is the same: Which memories does the body save as PhiAlpha and which does it leave behind? How are these choices made?

For the dying individual, PhiAlpha uses the brain's limbic system as a filter to choose which memories are most necessary for preservation of the person's individuality at death. The hippocampus and the amygdala's emotion control center would most likely perform much of this task. They are both centrally located in the center of the brain, which would be among the last of the brain's components to deteriorate from oxygen loss following cardiac arrest or from whatever physical trauma the body suffered.

What memories does PhiAlpha choose?

This question forces us to examine how our individual personalities are constructed. Let us consider some examples.

When we look in a mirror and see our reflection—our emotional state is affected every time we remember our reflection. We are either happy with what our reflected image represents to or we are not. This emotional response is registered in our amygdala. This is an example of **ultra-short-term** or **sensory memory**. It will fade away in time because its impact upon the neuron carrying it is very slight compared to all the other memories contained in the cortex.

Illustration No. 6 (below) is an image of the five senses from which we receive millions of bits of sensory data every second. Sensory or ultra-short-term memory can include everything we are seeing, to the feeling of clothing and jewelry we wear, to how warm or cold we feel, and the scent of whatever odors are being carried in the air or cold we feel, and the scent of whatever odors are being carried in the air.

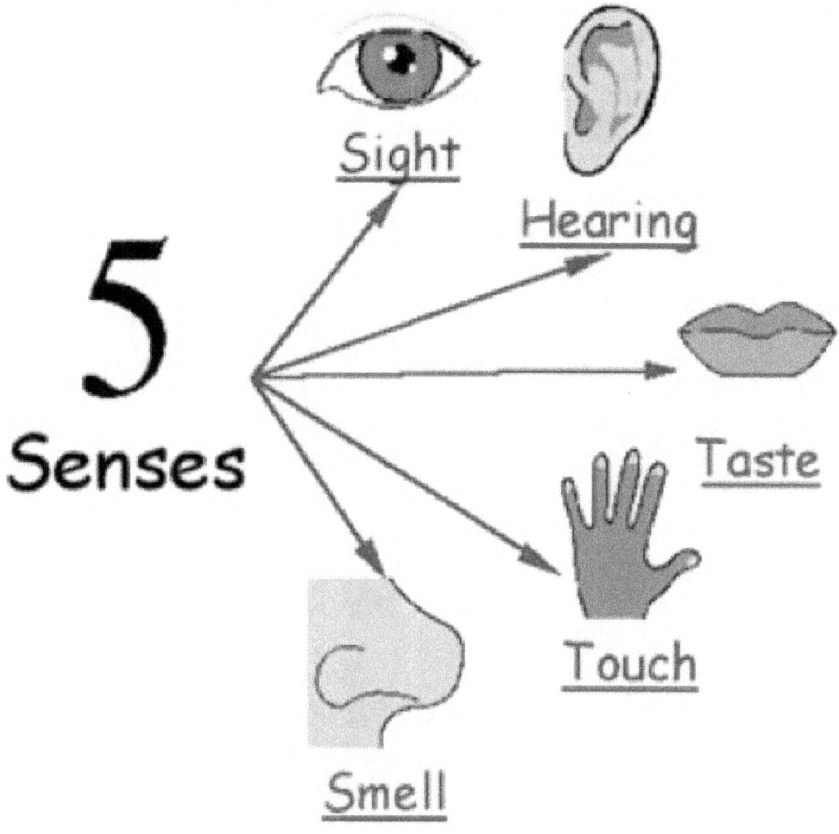

Illustration No. 6

Working memories do not last because they do not possess any emotional weight for neurons to latch onto. They are the flotsam of our day-to-day lives. They arise from the superficial contact of the conscious interaction we have with our external environment. When the drive to work is routine, there is no emotional hook upon which it can attach to a neuron. But we do remember the drive to work when we become irate over another motorist's carelessness.

The routine duties of the day can merge into one another. Each day has its own rhythm. Yet each day is so similar to every other that we take for granted most of what we do or observe. This is an example of working memory.

Illustration No. 7

Illustration No. 7 shows a typical example of what our working memory is responsible for routine day-to-day decisions such as our daily work, our daily schedules, and the subjective criteria we use to make the choices we make every day. The difference between sensory memory and working memory is based upon the question of which is more important to us: our physical environment or the activity in which we are engaged within that environment.

For example, the man in the photo above is seated at his desk where he is studying what is on his computer screen. If he is sitting on a wallet that is awkwardly placed in his back pocket, he will be uncomfortable. This is sensory memory. The report he is preparing, while enduring the increasing discomfort of that awkwardly placed wallet, is part of his working memory.

Long after he has forgotten his sore posterior, he will remember whether he was patted on the back for that report—or demoted. His brain's limbic system will register his emotional response to that evaluation on the neurons onto which the memory is encoded.

Both ultra-short-term and working memories would immediately be rejected for PhiAlpha. As the purpose of PhiAlpha is to preserve only those memories that support individuality, present sense impressions and day-to-day working memories are too immediate to carry much, if any, synaptic weight. Further, one of the means by which the hippocampus and amygdala prioritize memories is the strength and size of their neural-synaptic cells, which vary in strength and size depending upon how often they are used.

Long-term memories would logically be stored in neural-synaptic cells stronger and larger than recently created memories simply because they have existed over a longer period of time, and their synapses have been used considerably more often. This evaluation would require only a simple computational exercise by the limbic system to perform.

All that would be necessary would be for the limbic system to compare the strength and size of synapses of each long-term memory it receives. Upon cardiac arrest, ions containing long-term memories would be released into the neural net via the CSF wash. These ion-held memories would then be transferred to the amygdala for emotional prioritization by size-comparison.

A much more difficult task for the limbic system to perform would be to differentiate between long-term memories that are of equal or near-equal sizes. So, how does PhiAlpha select which long-term memories to save?

The strongest long-term memories are those memories we still recall, often without even trying. They bubble up out of subconscious like phantoms. Images of long-ago events washing over our minds like a motion picture we cannot turn off or walk away from. They seem to have a will of their own, dominating our attention regardless of what we are doing at the moment and keeping us up at night or waking us up from a sound sleep.

Evolution created PhiAlpha for the specific task of conserving information to comply with physical laws fundamental to the existence of the universe. Who each of us is as an individual is not something we can consciously identify or even observe. We are the object behind the face falling away from the canyon ledge into the mist far below—not the observer watching us fall.

Our conscious minds are held prisoner to the immediacy of our second-to-second sensory experience and the necessities of our day-to-day lives. Long-term memory creation is a program running far into our subconscious, if not farther.

If there is such a thing as a mystical component to PhiAlpha, it is at the point of long-term memory creation and prioritization where these two seemingly contradictory concepts meet. Our emotions do matter in the creation of stronger, longer-living memories.

Recent research supports this theory. It has been found that the brain's amygdala controls how the human brain preserves our emotional memories. The amygdala does this by encoding upon our memories the emotions we feel as we live through our life experiences. Emotion has both transient and long-term effects upon memory. Emotionally charged memories, including stimuli formed by pictures, words, and faces are more easily recalled over time that emotion-neutral memories.[73]

This research proves the emotional loading of our memories is controlled by the amygdala and corroborates a critical physiological step necessary for PhiAlpha's operation.

In 2003, a separate team of researchers studying the effect of negative emotions and long-term memory suggests negative emotions are the strongest and longest lasting of all memories—therefore, the easiest to recall. Further, long-term memories are given preferential treatment for recall over short-term memories.[74]

We can hypothesize that since negative memories are strongest, the synapses containing negative long-term memories have a stronger ion signature and are physically larger than other long-term memories. This preference for negative emotions in long-term memories is likely to be an evolutionary construct for survivability.

It is possible that evolution boosts the ionic signature of negative or life-threatening memories in the synapse as a means of self-preservation.

Gradual loss of brain oxygen at death limits the time for PhiAlpha to operate. All the processes necessary to download, prioritize, encode, and transmit memories must be completed before the limbic system exhausts its oxygen supply. Due to this time limitation, this process must be simpler than it might appear.

As the amygdala is the one brain organ capable of encoding emotion upon memory, it would be logical for it to possess an emotional matrix of the individual's personality—a blueprint, so to speak, of the person's identity over time (including from one life to another). This makes perfect sense given the fact that if PhiAlpha exists at all, there would have to be an organ within the brain capable of transferring emotions to memories and then encode both upon biophotons for emission out of the body.

If this is correct, all the amygdala would need to do to prioritize the ions from long-term memories is to run them over its own emotional matrix, holding on to those memories that match and disregarding those that do not. These remaining prioritized ion-held memories carrying the person's individuality-encoded memories would then be transferred to another area of the limbic system for biophoton encoding.

9. MEMORY ENCODING

Memory encoding requires three components:

1. The ions containing the memories

2. A place within the limbic system where the memories can be successfully encoded.

3. The transmission medium upon which the memories are to be emitted out of the body.

How the ion-held memories are likely prepared has already been discussed. Encoding of ion-held memories must be an electrochemical process. It may use the same principles of simple photography or something more complex such as magnetic resonance imaging (MRI). Whatever the process, it must be natural to the physiology of the brain and require little time and energy. A dying person would have little left of either. By this time, the gray matter and white matter of the brain may already be "dead" and the only part of the brain still alive may be the brain's core, where the limbic system is located.

The location within the limbic system responsible for encoding the memories for transport out of the body must be free of electrical activity that might create electrical interference with PhiAlpha encoding. Such an area would need to be structured such as to readily secure the memories in place for stable encoding. It would also need to be relatively free of cellular activity of its own—a relatively electrochemically-free zone within which the encoding could take place in the same way a darkroom is necessary to develop photographic film. Limited interference may be tolerable as cellular activity throughout the brain at this stage of the death process would most likely be greatly reduced or even absent.

One such area of the brain matching all of these requirements is the **hippocampal sulcus**; a small fissure or pocket that lies within the hippocampal formation near the amygdala. The hippocampal sulcus has the two qualities needed for PhiAlpha memory encoding.

First, its pocket-like shape can readily hold and stabilize the ion-held memories while they are being encoded upon the transport medium. Second, being relatively free of cells itself, the hippocampal sulcus provides an area relatively free of any electrochemical activity that might interfere with PhiAlpha encoding.[75]

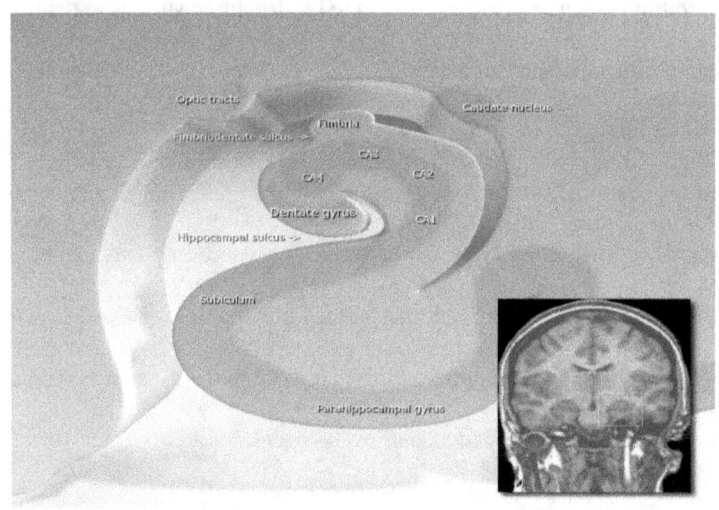

Illustration No. 8

How are memories encoded?

Illustration No. 8 shows a drawing of the hippocampal sulcus (the photographic inset is an MRI image of the brain with the hippocampal sulcus highlighted by the yellow square). As shown in the drawing, the hippocampal sulcus is shaped as an empty pocket-like fold located between the dentate gyrus and the subiculum.

The hippocampal sulcus is the brain's own little darkroom. The pocket inside the fold provides the compactness necessary to force the biophotons and the ion-held memories together. As the biophotons and memory-ions merge, there is a brief flash of light as the memories are bioelectrically transferred like a photograph from the ions to the biophotons. Once the memories are "flash-encoded" upon the biophotons, the biophotons are released and travel out of the body.

Ion-held memories prioritized in the amygdala would not have to travel far to reach the hippocampal sulcus as the hippocampus lies just below the amygdala. The ions would be held in place by the pocket-like formation of the sulcus long enough for the hippocampus to discharge a flash of biophotons from the surrounding sulcus.

This, we suspect, is the cause of the sudden, intense EEG burst observed by ER doctors treating patients who had near-death experiences. This flash of biophotons would extend throughout the surrounding hippocampal sulcus tissue, essentially creating a holographic image of the memories stored within the ions. This 360-degree flash of biophotons would have the added effect of encircling the ion-held memories in the self-contained state of energy and data coherence necessary for survival during emission out of the body.

Biophoton emission in humans has become an exciting area of study in recent years. A growing body of evidence suggests that photons play an important role in the basic functioning of cells. Cells emit light when they work and use light to communicate with each other. Photons may also play an important role in neural functioning. Spinal neurons of rats have been found to emit light. Microtubules inside the brain may act like optical fibers conducting photonic information.[76]

In addition to electrical and chemical signals propagating in the neurons of the brain, signal propagation also takes place in the form of biophotons. Experimental confirmation of photon-guiding properties of a single neuron have been documented by researchers. Interactions between biophotons and microtubules in the brain may cause transitions or fluctuations of the tubules between coherent and noncoherent states. The role played by biophotons inside the brain is an exciting new area of biophysics that merits special study.[77]

As Professor Majid Rahnama of Iran's Shahid Bahonar University suggests:

As Rhanama and his research team suggest:

"These interactions involve long-range ionic wave propagation along microtubule networks (MTNs) and AFs (action filaments) and exhibit subcellular control of ionic channel activity..."[78]

They explain how biophotons help support coherent energy states and show how the interaction of a quantized field with a system of electric dipoles produces what's called a state of coherent energy. They believe microtubules can be considered biological electrical dipoles.[79]

Other researchers have confirmed the presence of biophotons in both living and deceased human brains. Professor Joey M. Caswell and his team instructed human subjects to employ intention in effort to affect the direction of a "random number generator" within a special device located within 1 meter of the test subjects. With this device, biophoton emissions from the brain's right hemisphere were recorded simultaneously with each of the subjects' thoughts.

Caswell and his team discovered that when human test subjects attempted to affect the numbers produced by the "random number generator," their brains experienced a significant increase in photon emission. The researchers concluded that intentional thought patterns were directly responsible for these photon emissions. In other words, the researchers were able to conclude that that the human brain emits photons as a normal process of its functions.[80]

The interrelationship between the hippocampus and biophoton emission has been well documented by independent teams of scientists. Nicolas Rouleau, of the Biomolecular Science Program at Laurentian University in Sudbury, Ontario, Canada, measured biophoton counts that exceed background flux densities (electromagnetic field densities) from hippocampal tissue slices of and the brains of human test subjects engaging in specific thought processes.[81]

In a separate study, Rouleau found human hippocampal brain region specimens possessed the ability to emit biophotons 20 years after death.[82]

Lastly, a 2016 study found that glutamate-induced biophotonic activities and transmissions in the brain, present a spectral red shift from animals (in order: bullfrog, mouse, chicken, pig, and monkey) to humans, even up to the near infrared wavelength in the human brain. This increase in photonic activity in the human brain, they believe, may partly explain the high intelligence of humans because photons are a more efficient communicator than electrochemical processes.[83]

The subject of quantum mechanical processes within the brain has been an area of ongoing scientific debate for decades. But now that biophoton emission from the brain has been actually observed and studied, there can be no debate. It is proven. The brain emits biophotons continuously as part of its normal everyday function. The conclusion that the human brain is a continuous source of biophotons as the result of its electrochemical and electromagnetic processes is startling. Even more startling, however, is that this biophotonic behavior opens the door to the mind-blowing reality that the human brain is an organic quantum mechanical machine. Brain physiology creates a quantum mechanical energy state and uses this quantum energy to function in ways we have yet to even imagine.

10. MEMORY EMISSION

We are near the end of the PhiAlpha process within the dying human host. In our theoretical model, the individuality-carrying memories have now been encoded upon biophotons within the hippocampal sulcus. This packet of biophotons is now ready to be released or emitted out of the dying body. As we just discussed, biophoton emission out of the body would be controlled by quantum mechanical effects and may involve what researchers term "quantum teleportation."

In 1998, Professors A. Furusawa, J. L. Sorenson, and their team reported the first successful experimental quantum teleportation of photonic energy state.[84] In 2016, a separate research team reported successful teleportation of entangled photons over a 30-kilometer optical fiber quantum network.[85] Also in 2016, the team of Raju Valivarthi and Marcel.Li Grimau Puigibert reported the successful quantum teleportation of a telecommunication-wavelength photon interacting with another photon over 6.2 kilometers. Applications of this research in the telecommunications industry include extending quantum communication distances using quantum repeaters that rely upon light-matter (photon) entanglement.[86]

The concept of "quantum teleportation" is complicated, but its origin arises from a logical inference of quantum mechanics. The teleportation aspect of the concept is not what messes with our minds. Most of us are familiar with the concept of teleportation from *Star Trek*. It is the "quantum mechanical" component of the term that can throw us for a loop.

Quantum mechanics is a part of the universe humans lack the ability to experience with any of our five senses. We cannot see, smell, touch, hear, or communicate with the quantum mechanical part of the universe. This is so because the realm of quantum mechanics is limited to only those forces and matter so small that they are referred to as elementary particles. Elementary particles are so small they can only be studied by special machines called particle accelerators and decelerators, like the Large Hadron Collider and the Super Proton Synchrotron at CERN.

The first serious consideration of the subject of quantum-teleportation was the thought experiment Einstein, Podolsky, and

Rosen made famous in their 1935 article, "Can Quantum-Mechanical Description of Physical Reality be Complete?" Quantum entanglement is one of the weirder applications of quantum mechanics and theoretically creates, as Einstein described, "spooky action at a distance."[87]

To illustrate what Einstein meant, consider the following scenario: two particles interact and then both fly off in separate directions at the speed of light. Even when they are so far apart that classical mechanical laws no longer apply, a quantum mechanical measurement of either particle will instantly provide descriptive information about the other.[88]

This action is described as spooky because it implies that with entangled particles, information can travel instantly, regardless of the distance separating them—in apparent violation of the laws of classical mechanics stating nothing can travel faster than light itself (186,000 miles per second). There is nothing spookier than the idea that quantum entanglement allows light waves to travel across intergalactic distances instantly.[89]

Several questions pose themselves at this point regarding PhiAlpha emission. How much information can one photon carry? How many photons are needed to encode PhiAlpha memories? How many biophotons could be stored in the hippocampal sulcus and emitted together as discrete PhiAlpha before emission? How would this PhiAlpha matrix be emitted; as a collection of joined particles, or as a wave? Where would PhiAlpha be transmitted, and how would it get there?

These are all critical questions worthy of scientific interest. However, without an afterlife science, these important issues will remain unanswered indefinitely.

We should review some basic information. Photons are elementary particles that have zero-rest mass and are always traveling at the speed of light. Like electrons and other elementary particles, describing photon behavior (including biophotons) is restricted to the abstract logic of quantum mechanics. Photons are subject to what is termed "wave-particle duality." Photons share the properties of both particles as well as waves.[90]

However, the famous physicist and Nobel Laureate Richard Feynman believed both the wave viewpoint and the particle viewpoint were inaccurate. An accurate description of photonic particle and wave properties is very complex. While this subject extends far beyond the scope of this book, his lecture on the topic, archived at the California Institute of Technology, is well worth studying.[91]

We know from our earlier discussion of quantum entanglement that photons can be measured as entangled particles. We also know photons are measured in the electromagnetic spectrum as waves whose wavelengths vary with frequency or energy of the photon (which determines the colors we see in the visual spectrum). So, let us just accept as true that photons are both particles and waves relative to how we interact with them.

Measuring individual photons is extremely difficult. Like any elementary particle, photons are fragile and weak. Merely measuring them can be enough to interfere with their properties.

Photons of PhiAlpha would be emitted from the hippocampal sulcus either as discrete particles or as an electromagnetic wave or series of waves. They would most likely then absorb the ions containing PhiAlpha memories and immediately exit the dying or dead body as a wave of radiation with a specific frequency and wavelength. If correct, this emission of photonic radiation would then be measurable even with current technology. It would most likely be one sustained pulse that travels as a wave, though its frequency and wavelength are presently unknown.

A promising starting point for further research in this area should focus upon the correlation between the gyrus/sulcus fold-wave pattern and photonic wavelength. We do not believe it possible for neural net ions to travel from neuron to synapse to neuron through wavelike axonal channels without absorbing an energy signature matching the wave-path they traveled through the cortex by.

However, it is possible that the simple act of measuring PhiAlpha might be enough to destroy its energy coherence—in effect, destroying the memories it contains and the individual personality they are attempting to preserve. Even measuring elementary particles may be enough of an interference to affect their behavior. Therefore, it is important to create instrumentation that will be able to observe

PhiAlpha without adversely affecting the individuality matrix contained within.

11. MEMORY RECEPTION

So far, we have shown the steps needed for our individuality to be transferred out of our dying body: our individuality is downloaded in memory-signals from our neurons to the limbic system for prioritization. The memories are then transferred by ions to the hippocampal sulcus for flash encoding upon biophotons, forming a biophotonic packet containing the memories. The biophoton packet is then emitted out of the brain.

Where do biophotons go?

Light acts as both a particle and a wave. As a wave, PhiAlpha can be both observed and measured. PhiAlpha will have a wavelength and frequency just like any other electromagnetic wave. PhiAlpha will be capable of being received—just like any other wave—given the appropriate receiving antenna.

There are only two logical options as to what this receiving antenna could be. Either the receptor is a biological life-form or it is not. Simple logic dictates the conclusion that evolution intended for PhiAlpha to be received by the same species from which it was transmitted. Evolution is, by process and purpose, a biologically oriented process. Therefore, it can only transmit information from one life-form to another.

Evolution crafted PhiAlpha as an electromagnetic wave to be received and absorbed by a fetus in the womb as quickly as possible after emission and release from the dying host body. Any delay could result in PhiAlpha's energy coherence decaying, threatening the viability of the information stored inside. As PhiAlpha is most likely an electromagnetic wave, it is possible it may be intercepted, relayed to an intermediate storage area, and tampered with. This unnatural intrusion into the evolutionarily created and moderated procedure would threaten the very underpinnings of biological life itself.

We will restrict our discussion here to the most likely method of transmission: that the PhiAlpha matrix is transmitted as a quantum mechanical state and received by a human fetus through the antenna-like organ of the spinal cord before birth. This is not as strange as it may sound.

California State University, Sacramento, Professor S. Balaguru's paper on "Investigation of the spinal cord as a natural receptor antenna for incident EM waves and possible impact on the central nervous system" reported his team's research on the effects of EMF waves on biological systems including the spinal cord.[92] Their application of modern modeling technology known as finite difference time domain (FDTD) was used to simulate various voltages and electric currents in different parts of human and animal tissues. When they applied this technology to analyze incoming EMF waves to the spinal cord, they observed the spinal cord acts as a natural antenna.[93]

The spinal cord begins to develop with the rest of the brain at week five (three weeks post-conception).[94] Therefore, it appears that both the brain and spinal cord grow together from the beginning as one system. In fact, images of a 3-month-old fetus shows the spinal cord already developed as a fibrous filament that enters the brain and merges with the pons and cerebral peduncle as one undifferentiated tissue. At this stage, the spinal cord is shaped like a small thread or filament. The thread-shaped tissue of the spinal cord extends into the brain, curling like a snake around the thalamus into the midbrain itself.[95]

Illustration No. 9

Illustration No. 9 is a photograph of a complete human spinal cord and its connection to the brain. Special note should be given to the length of the cord and its antenna-like appearance.

Though the main part of the spinal cord is located within the spinal column down the back,

its nerve-roots extend like tendrils outward from the spinal cord. Together they create a web-like structure perfectly suited as a bio-antenna to receive the PhiAlpha wave signal.

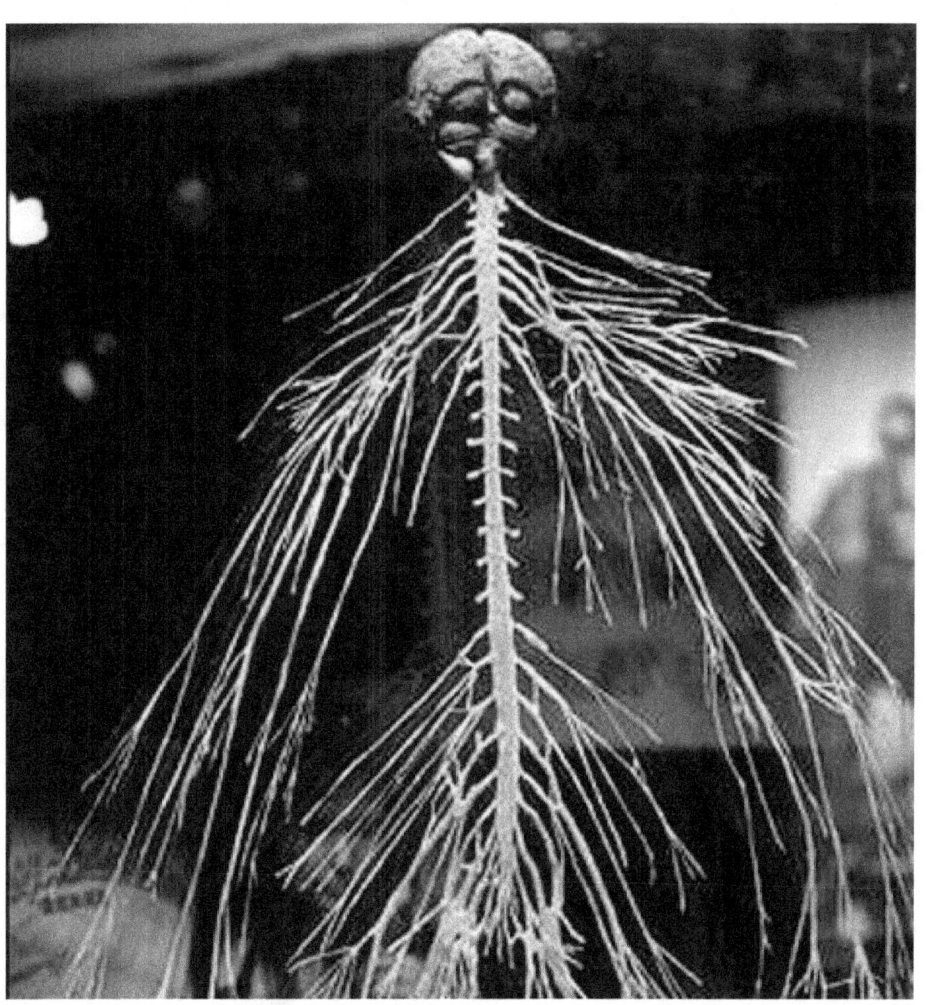
Illustration No. 10

Illustration No. 10 shows how the human spinal cord's nerve roots look when spread out, as they would be in the body. This illustration shows how similar the extended nerve roots are to many modern web-mesh antennae arrays used in telecommunications.

Like any other electromagnetic wave, PhiAlpha passes through the mother's tissue and is absorbed by the fetus's developing spinal cord like any antenna absorbs electromagnetic radiation of receivable wavelength. The size, length, width, and density of the cord and nerve-root tissue may all play a role in determining the wavelength and frequency required for reception of the PhiAlpha wave signal.

Once implanted within the tissue of the spinal cord, PhiAlpha memories are gradually unencoded from their carrier photons. The spinal cord tissue would separate the photonic particles from the ionic particles. The photons would be absorbed as energy, leaving the ions behind for decoding by the fetal limbic system when they were developed enough to perform that task. Until, then, however, the ionic memories would remain, stored safely within the spinal cord.

How long PhiAlpha can exist as a phased state of coherent, disembodied energy is unknown. In other words, we cannot say with certainty how long the emitted biophoton packet of individuality-encoded memories can remain viable without absorption into fetal spinal cord tissue. Without some device with which to artificially maintain congruity of the phased state coherence, we do not believe PhiAlpha could maintain the coherence of the photonically embedded memories indefinitely.

At first glance, the concept of our spinal cord tissue being an electromagnetic communication device sounds incredible. Nevertheless, the idea of a fetal spinal cord acting as a virtual radio wave antenna is not only possible, it is highly likely. Examine the following three illustrations as examples of biological and human-made antennae:

Illustration No. 11

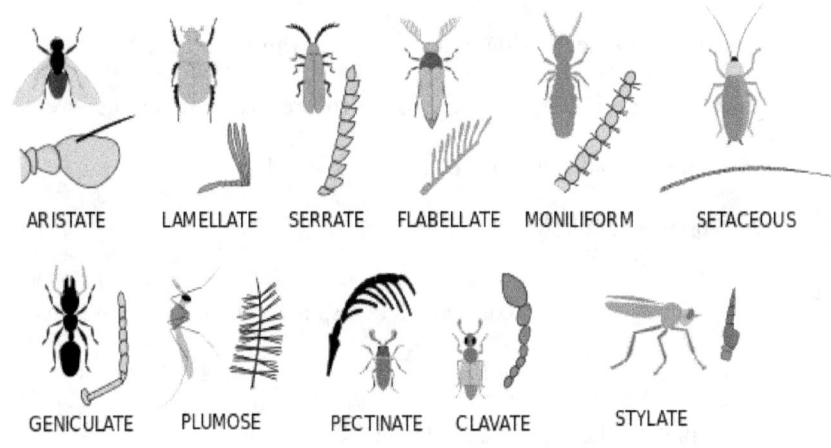

Illustration No. 12 *Illustration No. 13*

As the above illustrations demonstrate, biological and artificial antennae come in many shapes and sizes. In fact, many man-made antennae are modeled or copied from those of nature.

Illustration No. 14

Illustration No. 14 is a satellite voice antenna used by the United States military. Its wide frame mimics the web-like nerve roots depicted in Illustration No. 10. They both have a connecting center rod (in Illustration No. 10, this rod is the spinal cord) with horizontal extensions (in Illustration No. 10, these extensions are the nerve roots). While each looks different, the principle used is the same in each—reception of an electromagnetic signal.

12. VALIDATION TESTING

Validation testing is the technical term scientists use to describe the experimentation done to determine the legitimacy of any scientific theory. Why a chapter dedicated solely to this issue?

PhiAlpha is a mind-stretching theory. This is the first articulated description of a nonreligious alternative to the afterlife in human history. Sit back and ponder that a moment. How often do you have the opportunity to read and study something no one else has done—ever?

Now that you have had a moment of inspired contemplation, forget about it. Because there's a problem. The problem is proof. How do we go about proving PhiAlpha is real? All this book can do is point to the research that has already been done on unrelated topics and how that research paints a picture called PhiAlpha. All this book can do is point the way to what other research is possible to further explore whether the theory of PhiAlpha holds or not.

What follows is not meant to be an exhaustive list of the means by which we may now test PhiAlpha's validity. In fact, there is very little we know about many of the brain's organs—especially those within the limbic system. It may be years, even decades, before we are able to see the brain in its wholistic entirety. Until then, we continue to struggle for a picture of the forest while lost among the trees.

But to every scholar and scientist and researcher who doubts the existence of PhiAlpha, it is hereby demanded: before you argue against PhiAlpha's existence, first find evidence that it does not exist. For, in this book, is compelling evidence that it does.

There are at least three ways of researching the existence of PhiAlpha that do not involve active capture of a PhiAlpha packet: (1) passive monitoring of the dying for biophoton and/or ion emission; (2) passive monitoring of a fetus for biophoton and/or ion reception; and (3) genetic research.

Clearly, active capture of a PhiAlpha packet (which is not possible with present technology) of biophoton-encased ions carrying a human memory download at death would involve significant legal, ethical, and moral issues making such an effort untenable. Therefore, we must be extremely careful how we proceed. Passive-monitoring (which is possible with present technology) is clearly the safest approach and ensures no interference with the PhiAlpha packet during testing.

As we plan our experimentation program, we must consider what the possible impact upon the PhiAlpha packet would be if the theory is valid. How would the individuality preserved by the ion memories encased in the biophoton packet be affected by our testing techniques? As we monitor the subjects, we must always remember to place ourselves in their position and assume the responsibility as guardians of their preserved memories, which may be all that remains of their consciousness and individuality. It is, indeed, a most sacred trust.

In response to those who contend that with death the deceased no longer possesses any legal standing or right to complain about what happens to what is now a corpse, let us remember that our purpose here is to test the validity of a theory of possible immortality—indirect though it may be.

If one's memories are immortal, then is not the one from whom the memories originate immortal as well? Can we say it is impossible for one's future physical self to remember what was done to one's memories during the transition from dying host to fetal host? These are both reasons why passive monitoring for possible biophoton emission at death is the only acceptable validation regime.

It is not necessary to capture a PhiAlpha packet to validate the theoretical model. Technology exists permitting the passive monitoring of a dying subject's biophoton emission. Such a program would involve a human test subject dying in a controlled environment (hospital ICU) while researchers monitor electronic equipment, such as single-photon detectors attempting to observe one or more photons, photon packets or ions emitted from the dying subject. Detection of a photon or ion emission, under controlled conditions, would be compelling evidence that PhiAlpha exists.

The past few years have seen a rapid development of single-photon detectors (SPD) for passive monitoring including photon-number-resolving single-photon detectors such as: (1) transition-edge sensors (TES); (2) superconducting tunnel junction detectors (STJ); (3)

parallel-superconducting nanowire single-photon detectors; (4) charge-integration single-photon detectors; (5) visible light single-photon detectors; (6) quantum-dot optically-gated field effect transistor detectors; (7) time-multiplexed single-photon avalanche detectors; and (8) single-photon detector arrays.[96]

Photon-number-resolving SPDs are capable of precisely counting the number of photons emitted that makes this type of single-photon detector our preferred choice over detectors that do not count photons. We want to know the precise number of photons needed to encase the ions we hypothesize are encoded with memories at death. As you may recall, the photon-packet is necessary to ensure continued coherence of the ion-memory core as it leaves the dying body and transitions freely as an EMF wave to a receptive fetus.

This leads us to the many technical issues that will have to be resolved during testing. We do not know (and will not until we begin testing) how many biophotons are needed to transport PhiAlpha memories from a dying host to the fetal host. Nor do we know at what wavelength the biophoton(s) will be emitted. Will the emitted biophoton be in the ultraviolet? Will it be in the infrared? We do not know.

As a result, we will need a detector that can be calibrated for as broad a wavelength coverage as possible. If that is not practical, we will need to employ more than a one detector to ensure all possible electromagnetic wavelengths are monitored during testing. Certain wavelengths will create their own problems. For example, near-infrared and mid-infrared wavelengths require narrower nanowires (20-30 nanometers wide)[97] and lower superconducting critical temperatures.[98]

As challenging as the scientific and technical issues will be to resolve, they will not be as crucial to the project as will be those issues relating to the use of human subjects. We are convinced we now have the technology to test our theoretical model's validity and that all scientific and technical issues that arise during the program will ultimately be resolved.

The other two means of testing existence of PhiAlpha are:

(1)   attempting to measure a biophoton packet of memory ions being absorbed by a fetal spinal cord.

(2)   attempting to locate genetic markers of PhiAlpha through genetic research.

Each of these have their own advantages and disadvantages as practical research methodologies.

First, regarding fetal-spinal-cord absorption, we lack knowledge as to exactly when this occurs. Does it occur in the womb or after birth? Our preliminary hypothesis is that absorption probably occurs in the womb and that downloading memories from the spinal cord into the fetal neural net occurs gradually and may well continue into early childhood.

We hypothesize that fetal absorption of memories would serve as the "on-switch," jump-starting the process of increasingly complex associations within the fetal brain (ultimately resulting in a critical mass event of self-awareness we call consciousness). What we would look for is some evidence that the fetus is exhibiting planned behavior evidencing this growing perception of self.

Evidence of this growing awareness may be found in what some mothers report as fetal movement within the womb (called the quickening or fetal flutters). These movements often begin as early as 12 to 16 weeks from the start of the mother's last period. During early pregnancy, these movements are irregular and show no sign of intention.

However, at some point during the third trimester, these movements take on a clear regularity and can be timed by the mothers. This pattern is apparently unique to each unborn child, so mothers are often asked to time these movements to become aware of their baby's individual rhythm. Clearly, the regularity of the movements is indicative of intention and, therefore, evidence of self-awareness.

The problem is that to we do not know how much of a download of memories from spinal cord to neural net is required to initiate self-awareness. It could be immediate—or it could take weeks or even months. At this point, we just have no way of predicting the timeframe involved. Until this problem can be resolved, monitoring the expectant mother during the entire pregnancy will pose its own issues.

Our third option is genetic research. However, as we discussed earlier, PhiAlpha necessarily involves interaction between multiple biological systems in the human body, likely coordinated through the limbic system. These interactions may not be easily deduced or discovered genetically due to the indirect nature of PhiAlpha processes.

Further complicating detection, we believe, is that PhiAlpha may itself be a near-death upregulation of genetic markers that lie dormant during the rest of one's lifetime. If this is the case, we fear detecting genetic locales where PhiAlpha processes have awakened will be extremely difficult to detect until the precise moment of death.

As a result, we conclude that the testing program that provides the greatest opportunity for a scientifically valid result is the program of monitoring a subject who is near death. If pursued consistent with the concerns addressed herein, we believe such a program offers the best chance of humanely obtaining evidence of whether PhiAlpha exists.

The most critical factors of this program will be selection of suitable subject candidates and a test location. These two factors are interrelated. Due to the research involved, all test subjects must be individuals who are expected to die in the near future. Their care and comfort must be the most important consideration. Informed consent will be required for every candidate in addition to at least one, if not two, family members. No candidates should be accepted into the testing program (no matter how motivated they may be) unless they enjoy the full and active support of their families.

Candidate evaluation and selection is, therefore, of critical importance. The testing program requires that nonfamily professionals be present at a person's death—the very time families want privacy and seclusion. Therefore, consideration must be given, not only to the physical, emotional, and spiritual well-being of the candidate but also that of his or her family. Resistance or uncertainty by either candidate or family concerning the test process should disqualify that candidate from participation in the program.

Test candidacy is a decision, not just for the potential subject but for his or her family as well. Written consent must be obtained from family as well as the candidate. Candidates and their families will also need to be notified of all known possible risks, including that their biophotons may be absorbed by the electronic monitoring devices during monitoring. The risk of electron absorption during testing has been noted in recent literature.[99]

To ensure candidates are mentally competent to agree to be a test subject, each should be given a thorough psychological/psychiatric evaluation at the beginning of the evaluation program with regular follow-up evaluations as time of death nears to ensure the subject retains competency to withdraw his or her consent until the end. Any evidence of mental incompetency should immediately disqualify the individual as a test subject. In addition, no one should be accepted as a test subject who has an appointed legal guardian, guardian for medical decisions, or has allowed anyone to have power of attorney authority. Any of these should automatically disqualify a candidate.

Mental competency and informed consent laws vary from state to state. Care will need to be taken to ensure compliance with all applicable statutes and governmental rules promulgated to regulate these issues. This is another reason why universities are the preferred structure for this program. Most major research universities have legal departments experienced in handling the complex human-test-subject legal and ethical compliance issues. These universities also have research hospitals capable of facilitating all necessary program requirements. Finally, it makes logistical and organizational sense to employ a university whose physics department and research hospital can coordinate the actual testing.

As candidate and family interviews, evaluation, and preparation will take substantial time, it will be necessary to select as potential test-subjects individuals who fit the following pre-screening medical criteria: (1) chronically ill for over a year; (2) terminal diagnosis at least six months prior to screening; (3) no longer expecting a cure and resigned to death in the near future; and (4) a desire to share one's last hours for scientific research for the betterment of humanity.

Can we expect an abundance of willing candidates? There will be those for whom such a sacrifice of privacy at so delicate a time will be too much to accept. However, due to the historic scientific and social significance of the research, suitable subjects may not be difficult to find.

There will be many more who see being a test subject as an opportunity to do something great and noble in their last hours. They would have the comforting knowledge that their act of selflessness helped humanity grow in self-awareness.

13. CONCLUSIONS

PhiAlpha is a theoretical model of how evolution may have responded to forces of nature compelling it to conserve information. We have described PhiAlpha's stages, from evolutionary origin through each stage of its process, and have explained how these separate stages come together to operate as PhiAlpha. Lastly, we have provided the reader cited references establishing each stage of PhiAlpha as scientifically valid. Every stage of the PhiAlpha process has been described, explained and its application shown to be both biologically valid and corroborated by current science.

PhiAlpha is the first articulated theory of how evolution most likely adapted human biology to attain a form of immortality through the preservation of individuality-encoded memories. PhiAlpha is presented to encourage academic and public debate upon the possibility of immortality through evolution. This book is also dedicated to the need for a serious, scientific analysis of the potential for life after death.

This book does not prove PhiAlpha exists. But, from what we have observed through the analysis of our theoretical model, neither can we say PhiAlpha does not exist. In fact, this book presents sufficient evidence of biochemical and quantum mechanical feasibility in PhiAlpha processes that we must conclude there is a probability that PhiAlpha does exist.

The most compelling argument on behalf of PhiAlpha's existence is that it just makes pragmatic sense. Evolution is built upon survivability and heritability. If a need arose to require survivability of a genetic trait such as memory, evolution would create a means of doing so and protect it from undue tampering or interference—just as it created gyri and sulci to increase cognitive awareness.

When examining the merits of a criminal case, trial lawyers analyze the accumulated evidence for what we trial lawyers term "reasonable suspicion" and "probable cause." These terms are expressions we, in the legal community, use to measure probability a crime was committed, and the suspect was the one who committed the offense.

As the term implies, "reasonable suspicion" suggests a conclusion that sufficient evidence exists to require further investigation, such as a voluntary encounter, stop and talk, or even a consensual search or voluntary interrogation. These are preliminary methods law enforcement utilizes to gather more evidence against the suspect. These methods may or may not yield further evidence.

However, if more evidence of guilt is obtained, the level of certainty that the suspect may have committed the offense in question may rise to the level of "probable cause." A conclusion of "probable cause" is a legal conclusion made by a judge, prosecutor, or law enforcement that the probability a criminal act perpetuated by the suspect did occur is greater than 50 percent, or, to put it another way, the odds a crime occurred are "more likely than not." In a court of law, probable cause is sufficient to justify arrest and prosecution.

Where should we place PhiAlpha on the gradient of scientific certainty at this point? At this point, PhiAlpha is but an untested theory. Yet, as this paper amply provides, there exists enough evidence of PhiAlpha to conclude there is reasonable suspicion that some form of biological process is at work that operates at death to preserve information, enough that testing should begin in earnest as soon as possible.

With PhiAlpha, death is not an end of life—and never was. With PhiAlpha, energy and information are conserved through transformation via biochemical, thermodynamic, and quantum mechanical processes allowing our individuality to survive our "physical" death.

We cannot say whether our human host bodies were ever meant to be immortal. Based upon our discussion here, however, we can say our individuality (personality) was meant to be as immortal as the laws of conservation of energy and entropy (information) allow. The only limitation would be the sufficiency of human host bodies (now being reproduced at will). As long as humanity survives, there is continuity of individuality—through PhiAlpha's preservation of memory.

Research into PhiAlpha's validity can be accomplished without risking capture or other interference with PhiAlpha particles. We recommend no one attempt to measure or capture PhiAlpha particles until these issues are thoroughly investigated and there is consensus that such efforts are harmless to the identity being transported. PhiAlpha is life at its purest but also at its most vulnerable and evanescent. It is a sacred life-form. No one has the right to interfere with its existence.

There will be those who fear PhiAlpha threatens their religious beliefs. Evolution has been developing PhiAlpha long before we humans were created. We are its progeny whether we like it or not. What force or forces set evolution in motion has us firmly in the palm of its hand. Our approval does not appear to be necessary or even requested.

However, all religions share common values and concepts that are also inherent within PhiAlpha: such as (1) karma; (2) thoughts are things; (3) what we do to others will come back to us tenfold (in memories and their impact upon our individuality); and (4) the kingdom of heaven lies within us. Each of these is consistent with the best of every major religion and PhiAlpha. Perhaps humanity's embrace of spiritual meaning is but a subconscious echo of the PhiAlpha within each of us.

PhiAlpha is an instrumental component of life. PhiAlpha is the transfer of individuality from one life-form to another. PhiAlpha has a sacred, spiritual context independent of our religious ideologies. It is quite possible our ancient ancestors were instinctively aware of PhiAlpha but gave it a religious label out of ignorance and/or misplaced allegiance to their priesthood.

Life is a gift beholden to no one and, therefore, sacred to all. Religion sees science as a threat to its power base and rightly so. Spirituality, however, has no reason to fear science. In fact, the two are mutually reinforcing. As Carl Sagan expressed so eloquently:

"Science is not only compatible with spirituality; it is a profound source of spirituality. When we recognize our place in an immensity of light-years and in the passage of ages, when we grasp the intricacy, beauty, and subtlety of life, then that soaring feeling, that sense of elation and humility combined, is surely spiritual. ...The notion that science and spirituality are somehow mutually exclusive does a disservice to both."[100]

Carl Sagan (1997)

Since our earliest humanoid ancestors stared up into the heavens looking for their celestial reflection, the universe has bestowed upon them and their progeny an infinite variety of wondrous gifts and mysteries, each one yielding its singular piece of the puzzle that is the fabric of the cloth we call existence. PhiAlpha presents one more piece of fabric for us to wear.

14. PHIALPHA

Given the evidence above, we can now assume that PhiAlpha is valid. Let us also assume that PhiAlpha's emergence is the result of evolution's adaptation of human biology to conserve what is uniquely human—the ever-expanding trait of individuality.

The following conclusions then become irrefutable:

1. Since the dawn of humanity, PhiAlpha has been present within our biological processes, preserving the memories that define us as individuals and then passing them on to each successive generation. We are the children of those memories. Our lives reflect the personalities—both good and bad—developed over those many lifetimes.

2. We carry with us from lifetime to lifetime a memory matrix of all our past lives. This can be nothing else but what is referred to as the soul. This soul is not mythical but a physical structure, however small, which must obey natural laws. The superficial projection of the soul upon physical reality is the self we are consciously aware of

during day-to-day life. In other words, the self is the reflection we see when we stare deep into the impenetrable depths of our souls.

These assumptions are not inconsistent. PhiAlpha lends physical validity to the existence of a soul and its existence as an eternal storehouse of individuality consistent with the law of conservation of information within the second law of thermodynamics. We can see the soul for what it is, not as a mystery religion cannot explain but as a physical reality—as memory-encoded biophotons that transfer human individuality from one life to the next. Thus, the human soul and PhiAlpha are intrinsically intertwined. They are the same.

What is "Spirit" if not photons from the Big Bang. They are the carrier-wave of the soul. "Spirit" (the Big Bang's photonic wave) is the coherent energy matrix that binds each soul together with one another and as part of the universe.

The soul is the eternal essence of human individuality which we create from life to life. PhiAlpha is simply the biological process that evolution has crafted to ensure continuity of the soul between each life.

Spirit exists as vibrating waves of universally harmonious energies that fill the soul with knowledge, love, compassion, inspiration, and creation. The soul is a transcendent instrument of spiritual enlightenment which reacts to this spirit. Each of us has been given a soul to play like a piano as we compose symphonies of living music with each of our lives.

What makes one soul different from another? The answer is that each soul becomes increasingly unique from each of the others by the memories each accumulates during the multiple lifetimes each one experiences. Each lifetime adds a new layer of memories upon those that came before. Each lifetime further differentiates each soul from the others.

PhiAlpha does not exist to destroy religion. PhiAlpha was not created to attack religion. Evolution and PhiAlpha existed long before and independently from religion. It is not PhiAlpha's fault that it conflicts with Christian teaching. It is the fault of Christian "leaders" who chose to build their "temples" upon ideological sand.

PhiAlpha cannot be prayed away or excommunicated into nonexistence. PhiAlpha has always existed—it exists now, and it shall exist into infinity—despite any resistance from Christianity or any other religion.

While PhiAlpha benefits from an internally consistent logical flow, it shall remain only as a starting point for actual scientific research. It is only a theory. Future discoveries may reveal PhiAlpha to be more, or less, correct than other theories that will eventually be developed. What is important, however, is not to let this beginning become lost in the noise of a substantial religious backlash.

Some will argue that PhiAlpha is a threat to religious doctrine in general and Christianity in particular. They will claim that PhiAlpha challenges long-held biblical beliefs, among them the existence of a literal heaven and hell; that Christians should not believe in any form of reincarnation; whether one goes to heaven or hell depending upon baptism, confession of sin, and faith in Jesus Christ as savior.

It will be very difficult for fundamentalist Christians to adapt to a world in which life extends beyond the grave through PhiAlpha—an evolution-originated process. Their religious indoctrination has turned their minds into an intellectual prison—nothing new can enter, and they cannot escape. Until something radical and profound enough comes along to shatter those concrete bonds, they are lost.

PhiAlpha may be just what they need. One scientist from a well-known religiously affiliated American university made the startlingly unscientific claim that PhiAlpha is not science because it can never be proven. This scientist's emotional reaction reveals the extent to which PhiAlpha threatened his religious-ideological foundation.

This is what happens when an otherwise intelligent, highly educated person allows his religious indoctrination to overwhelm his scientific discipline. No true scientist worthy of the profession would ever use the word "never" concerning a theory. Technology's continuing advancement has proved the point over and over again. Today's unproven theory is often tomorrow's reality.

If this "academic" thought about what he was saying, he would be ashamed of himself. Until recently, gravity waves were purely theoretical. We lacked the technology to prove they exited. But I do not recall any academic claiming the study of gravity was a waste of time. Until the mid-1900s, we could not prove $\mathbf{E = mc^2}$, but that did not stop American and German physicists from trying to build atomic bombs based on only Einstein's theoretical computations.

So, we must live with PhiAlpha, whether we like what it implies or not. We cannot not make evolution go away simply because its implications are inconvenient or because it challenges our preconceived notions of religious doctrine. PhiAlpha, like any other theoretical model is based upon what logical inference suggests may be possible—not what is obvious or necessarily apparent.

Despite Carl Sagan's quote in the previous chapter arguing that science and religion can coexist, there are those for whom every new scientific revelation is but a threat upon their cherished religious beliefs.

How will the concept of human immortality through evolution (PhiAlpha) be perceived by Christians? For that matter, how will PhilAlpha be perceived by Jews, Muslims, or Hindus? This will, of course, depend upon a combination of one's own religious indoctrination, how one's life experiences have shaped those initial teachings, and the religious community with which one is affiliated.

We are still only in the beginning stage of humanity's spiritual enlightenment and, as a result, we can expect a variety of reactions across the entire spectrum of understanding. Of course, the more one tends to believe in a strict orthodoxy-based religious belief system, the more likely one will object. This is only logical, as it is fundamental religious belief that will be at greatest risk from a unreligiously controlled path to the afterlife.

As understanding of the evolutionary and biological of PhiAlpha grows, the more fundamental and conservative Christians may see PhiAlpha as a threat to the relevance of their religious beliefs. As Americans increasingly seek to merge their religious views with political parties and candidates for elected office, their resistance will likely become politicized—just as evolution has become.

This resistance, a result of fear and ignorance, would be both unfortunate and unnecessary. True Christians and those who are sincere in their spiritual humility regardless of their religious upbringing and indoctrination, will realize biologically based immortality does not mean the death of their God. They will understand the underlying premise upon which religion must be judged:

God must be separate and apart from its creation—and act independently. The gardener is not the garden, and the garden does not define who the gardener is. The garden is merely a reflection of the gardener, nothing more. Therefore, humanity is not limited to the essence of God alone, but is free to grow beyond its design—and its designer.

But only if individuals choose to do so. Are we willing to accept that responsibility?

The concepts of godliness that we have developed within our own minds and shared with each other are not dependent upon the man-made rules of religious organizations whose primary function is to make money and wield political and financial power and influence for the sake of their leaders' material wealth and egos.

For all we know, genetics is a gift from God. If true, then God also created PhiAlpha. With PhiAlpha, creationism can no longer exist without science any more than it can live without God. If God created the universe, it also created the physical and biological laws that rule its processes. They are inseparable. Any other conclusion is illogical and irrational. PhiAlpha permits no other conclusion.

Remember what PhiAlpha does. PhiAlpha preserves the unique information that forms the basis of our personalities and individuality by creating a safe harbor for the memory-encoded biophotons of the dying until absorbed in the tissue of a fetus.

Wouldn't God want us to be individuals? Didn't God grant humans free will? If God did not want us to be individual aspects of creation, then God would not have created individualized souls. Further, if God respects the concept of free will then it has no choice but to expect us to develop our own individuality and respect the consequences that free will permits—both good and bad. There is nothing ungodly in that.

If you are a gardener and you plant an apple tree, you must expect it to grow apples. If you plant the apple tree, you accept that you will get both good apples and bad. You leave the tree alone. You don't kill the tree off just because it produces some bad fruit or because some of its branches grow in unpopular shapes and sizes. If you don't want to deal with bad apples, don't plant an apple tree.

Does anyone contest the concept that our souls were each meant to be unique from one another? Is there anything more sacred to Christians than the precept that each soul is unique to its host? Is not one's soul truly one's own? Can I possess your soul? Can you possess mine?

The answer is obviously no. Each soul can only be owned by an individual human. So, the concept of the soul, itself, must be based upon the creation of spiritual individuality manifested by the human spirit and body. This tenet of soul individuality is one that Christians worship as a strongly held belief of their religious doctrine. To believe otherwise would be to allow one person's actions in life to determine the spiritual fate of another.

Individuality of the human soul is a gift, not a curse. True, the attainment and development of one's own individuality is one of the most difficult processes we humans face during our lives—but also the most rewarding.

Most see their own individuality as a priceless artifact to be treasured, nurtured, and protected at all costs. For every person I have met who expressed frustration over their interaction with others and/or the world, I have met a thousand who loved who they were as individuals. Though, for some of them, it was the rest of the world that was at fault.

So, any god creating a process that preordained the origination of individuality would logically expect the beings borne with this capacity to be individuals who would cherish the gift. The soul, whose very nature is individualistic, would not wish to live its life in the pursuit of its blossoming individuality only to lose what it had gained, at such a high cost, just because its host, the physical body, died.

Humanity has evolved over billions of years from one-celled organisms to what we are today. That is a long time to wait for a tree to grow—even for a god. If individuality is the main fruit of this Tree of Life, would God allow its value to be wasted simply due to the transiency of human life? That would be internally inconsistent, and internal inconsistency cannot logically exist within the same reference frame of any given system. No system can pick and choose which laws to obey and when to obey them. Continuity of expression is a fundamental precept of our universe's natural laws.

The concept of PhiAlpha goes beyond the conservation of unique information required by the second law of thermodynamics. The science of PhiAlpha is founded upon an inherent logical consistency. There is no logical purpose to the origin and development of individuality without provision for preserving its end product—individuality's ever-increasing complexity of expression.

Religious fundamentalists believe creation took six days. If creation was completed within the first six days of existence, what has been happening ever since? To believe creation ended at the Big Bang is to believe the first chapter of a suspense novel is the climax, or that a Broadway play ends after the first scene or that a football game is over right after the kickoff.

We humans tend to jump to conclusions without having enough data. We think we know all the answers, but often we don't. So, we make mistakes that could have been avoided.

Creation of individuality without the provision for continuity would be like creating the laws of physics but denying the universe the physical ability to form planets, solar systems, and galaxies consistent with those laws. The consequences would be internally inconsistent with the causal processes. This would be neither logical nor rational.

Religious faith is not a logical process, however. People who act upon their religious faith do not do so as a result of what they observe or perceive as objective reality but often in spite of it. Their actions are the result of their inner voice and the subjective emotions that arise out of it. Conscience exists independently of external forces. The force within has greater influence upon behavior than any external stimulus. The more one's life is dependent upon one's religious faith, the more likely he or she will feel threatened by PhilAlpha.

The reality is that our religious growth is being stunted by the institutions that depend upon our religious obedience for money and power—the church. One need not be Jewish or Muslim or Buddhist or Hindu or Christian to understand the clergy could not survive without the faithful.

Therefore, I expect the most aggressive antagonists against PhiAlpha will be the clergy, the clerics, the priests, and the rabbis. It is, after all, their geese that PhiAlpha cooks. It is their careers and pensions that PhiAlpha imperils.

PhiAlpha ultimately stands for the independence of our human existence from any religion. With PhiAlpha, religions have lost their power and control over whether we exist after our body dies. It is not that God is dead—but that religion is. Religion is now irrelevant to the issue of an afterlife and whether we "earn it" by either our faith or good works. They aren't relevant any longer.

PhiAlpha reveals what the priests and clerics knew all along but refused to tell us: the only motivation for decency and morality in the world is what we, as individuals, resolve on our own to follow.

Living together in peace is a choice one makes regardless of religious affiliation. It is not a matter of faith that we seek peace but common benefit. We each survive. Civilization depends more upon the unspoken agreements between individuals than it does our collective religious faiths.

15. THE NEXT STEP

PhiAlpha's existence challenges us to cast aside what we thought we knew about life and death. PhiAlpha challenges us to reconsider what we experience every day as free will or predestination. Is creation static and unchanging? Or is creation an evolutionary process—an endless spiral of infinite realization?

As an example, let us take the issue of whether God created the heavens and the earth and humans in six days as it is written in the Bible's Book of Genesis. We are taught to believe that the universe must, therefore, be complete. Further, because "man" is part of the universe, we are taught that man is complete—but by sin, flawed.

With this doctrine, we are the equivalent of spiritually indentured servants who are given only one path to freedom (heaven)—through the church. To earn a place in an afterlife of heaven, we must first kiss the ring of religious doctrine; be it papal or ISIS or Taliban.

However, PhiAlpha tells us a much different story. PhiAlpha tells us that creation is an ongoing process during which nothing is fixed and all can become enlightened. PhiAlpha tells us the Big Bang was just the first note of an eternal symphony of creation and self-realization.

The universe is in the process of creation every moment of every day. The force or forces that created evolution are of life. Inherent within life is the freedom to adapt and change to survive and flourish—both spiritually and physically.

The galaxies did not just appear and then stop in place at the Big Bang. No, the galaxies, their stars, planets, and the beings living on them have been in motion ever since. creation is as infinite in time and space as the universe itself. Thanks to former Bell Lab's Robert Wilson and Arno Penzias, we know the echo of the Big Bang can be heard, even today, as cosmic background radiation. creation has not ended. In truth, the human chapter of creation has just begun.

Consider, for example, the idea that since we have not yet been in knowing contact with an alien civilization, that we must be alone in the universe. This is akin to saying the world is flat.

It took primordial life on our Earth billions of years to develop—and billions of years longer to evolve into the sentient beings we have become. Are we that much different from the rest of the cosmos? If it took life that long to develop here, what says alien races evolved faster? Where is the evidence it took less time anywhere else?

There isn't any such evidence. We have yet to discover life beyond this planet, let alone advanced civilizations. Therefore, for now at least, we here on Earth are the best evidence of how universal creation and life interact.

PhiAlpha takes us one step further. PhiAlpha says our path to a new life is no longer dependent upon religious doctrine at all. PhiAlpha shows that we have infinite lifetimes—not because we are good or bad, saved or unsaved—but because our spirits are eternal. If you can get to the next life without believing in a god—what is the point of the clergy or the church? What other product can they sell us that we would buy?

What is the next step?

If PhiAlpha makes present Judeo, Christian and Islamic doctrines irrelevant, does this mean the church itself is now irrelevant? Like any other human organization, the church suffers its share of internal political thuggery and mischief. The organization of the church will never be perfect because it will always be subservient to imperfect mortals.

PhiAlpha teaches that we are in control of our own individual spiritual destiny. We have always had the power to do so. The church simply hid the truth from us. PhiAlpha tells us we must take over our own spiritual destiny. But first, we must accept the mantle of authority to judge ourselves and each other as part of our communal oneness within the spirit of creation of which PhiAlpha is a part. This is the lesson of self-accountability and redemption through mutual forgiveness and good works the world needs today.

PhiAlpha teaches that we must stand on our own two feet spiritually. We must be willing to accept responsibility for ourselves and our communal (local, state, and national) behavior in the world. We must become our brother's keeper, inspiring each other to seek greater meaning and personal enlightenment while also demanding individual accountability at all levels of society.

PhiAlpha is the beginning of our spiritual renaissance, but it is not the end—far from it. Humanity has yet to figure out the whole "spiritual deal"—just as it has yet to figure out many of the countless mysteries of the universe.

What is the next step?

The existence of PhiAlpha demonstrates that our DNA is constructed to allow our consciousness to become more complex with each generation. Consciousness is a commingling of classical and quantum mechanical processes affecting our intellect and emotion, our physiology and spirit, and our rational and subjective perceptions. All this information is integrated within our individuality through PhiAlpha.

Does PhiAlpha recognize the existence of a soul? Of course, it does. But let's be careful as to how we define "soul." The soul PhiAlpha recognizes is that part of the human mind that is emitted within the biophoton capsule as the body dies. This is the part of us that survives death and passes on our individuality to a new generation of physical bodies.

PhiAlpha also establishes that each individual soul was created to become increasingly enlightened with each rebirth. The soul is an integral part of our consciousness. The raising of one's consciousness enlightens both the intellect and the spirit as well. The biophoton capsule of memories carried over from our past life becomes the seed out of which our individuality will be resurrected in the next life.

Enlightenment is impossible, however, until the mind is freed from the self-imposed chains of limitation and the external chains of doctrinal intolerance. While some explore enlightenment on their own, most remain spiritually bound by religious doctrines that are 1,700 years old.

Concepts that were created to control the interaction of simple merchants, hunters, fishermen, warriors, and farmers have little meaning to us now. And, do not forget the way religious doctrines were used to control women. Women were regarded as little more than property—and often much worse.

The power of mind control cannot be doubted. Two thousand years have gone by, and these same institutions still control the behavior of millions in the world today.

That must change. Enlightenment of the intellect and the spirit must be more than a hobby for the bored intellectual elite of the world. Unless freedom from mind control is made available to the masses, the gains made will be slow and only with grudging resistance.

What is the next step?

The next step is liberation of the human soul. PhiAlpha is the beginning of a new reality, a new vision of humanity's purpose in the universe. With PhiAlpha our innate spirituality can rise out of our subconscious, free of the artificial constraints of religious myths, to finally stand as inheritors of our divine purpose.

REFERENCES

1. Craffert, P.F., "Do out-of-body and near-death experiences point towards the reality of non-local consciousness? A critical evaluation," *Journal for Transdisciplinary Research in South Africa*, 2015, Retrieved at dspace.nwc.ac.za.

2. Borjifin, J., Lee, U., et al, "Surge of neurophysiological coherence and connectivity in the dying brain," Proceedings of the National Academy of Science (PNAS), 2013, Retrieved at cbi.nlm.nih.gov.

3. Chawla, L. S., "Surges of electroencephalogram activity at the time of death: A case series," 2011, *Toward a Science of Consciousness: Brain, Mind, Reality*.

4. Lake, James H., "The near-death experience: A testable neural model," 2016, *Psychology of Theory, Research and Practice, 4(1), 115-134*.

5. "All about entropy, the laws of thermodynamics and order from disorder," 2001, *Archives of Science*, Retrieved at entropylaw.com.

6. id.

7. id.

8. "What is thermodynamics," NASA Glenn Research Center, (n.d.), Retrieved at grc.nasa.gov.

9. Zhang, B. and Cai, Q., et al, "Information Conservation is fundamental: recovering lost information in Hawking radiation," Retrieved at arVix, 2013.

10. Horodecki, M., Horodecki, R., et al, "No-deleting and no-cloning principles as consequences of conservation of quantum information," arXiv, 2003.

11. id.

12. id.

13. Roncaglia, M., "On the Conservation of Information in Quantum Physics," arXiv, 2017.

14. Samal, J. R., Pati, A. K., and Kumar, A., "Experimental Test of the Quantum No-Hiding Theorem," *Physical Review Letters*, 106, 2011.

15. Shi, S., Kumar, P., and Lee, K.F., "Generation of photonic entanglement in green fluorescent proteins," *Nature Communication*, Dec. 5, 2017.

16. Tentrup, T., et al, "Transmitting more than 10 bit with a single photon," arXiv, 2017.

17. "Information in a Photon," (no author), DARPA (Defense Advanced Research Projects Agency, (no date).

18. id.

19. Noble, P.A., Pozhitkov, A.E., et al, "Thanatotranscriptome: genes actively expressed after organismal death," 2016, *Open Biology*, Retrieved at biorxiv.org.

20. id.

21. Jacobs, G., "An evolutionary view of individuality," 2012, *World Academy of Art and Science,* Retrieved at worldacademy.org.

22. Smith, John Maynard and Szathmary, Eors, *The Major Transitions in Evolution*, 1995, Oxford University Press, Oxford, England.

23. West, S.A. and Fisher, R.M., et al, "Major evolutionary transitions in individuality," 2015, PNAS, 112(33), 10112-10119, Retrieved at ncbi.nlm.nih.gov.

24. Michod, R.E., "Evolution of individuality," 1998, *Journal of Evolutionary Biology*, 225-227; See also Michod's many other articles regarding the evolution of individuality including

"Evolution of individuality during the transition from unicellular to multicellular life," 2007, PNAS, 8613-8618.

25. id.

26. McLean, C.Y. and Reno, P.L., et al, "Human specific loss of regulatory DNA and the evolution of human-specific traits," 2011, *Nature*, Retrieved at nature.com

27. Marino, L., Dunbar, R., Lahn, B., and Preuss, T., all cited in: Bradbury, J., "Molecular insights into human brain evolution," 2005, *PLOS Biology*, Retrieved at ncbi.nlm.nih.gov.

28. id.

29. id.

30. id.

31. Sherwood, C., et al, "A natural history of the human mind: tracing evolution changes in the brain and cognition," 2008, *Journal of Anatomy*, Retrieved at ncbi.nlm.nih.gov.

32. id.

33. White Matter. (n.d.) In Wikipedia.

34. Delude, C., "Researchers show that memories reside in specific brain cells," 2012, *MIT News*, Retrieved at news.mit.edu.

35. Stufflebeam, R., "Neurons, synapses, action potentials, and neurotransmission," 2008, *Consortium Cognitive Science Instruction*, Retrieved at mind.ilstu.edu.

36. Harris, K. and Sejnowski, T., et al, "Nanoconnectomic upperbound on the variability of synaptic plasticity," 2015, *eLife*, Retrieved at elifesciences.org.

37. id.

38. Neurons. (n.d.) In Wikipedia.

39. Chudler, Eric H., "Neuroscience for kids," 2016, University of Washington Press, Retrieved at faculty.washington.edu.

40. Neurons. (n.d.) In Wikipedia.

41. The brain core. (n.d.), The Brainwaves Center, Retrieved at brainwaves.com.

42. No author, "Amygdala and Hippocampus," In Wikipedia, date.

43. No author, "Stria Terminalis," In Wikipedia, no date.

44. No author, "Congulate Gyrus," In Wikipedia, no date.

45. No author, "Corpus Callosum," In Wikipedia, no date.

46. No author, "Fornix," In Wikipedia, no date.

47. No author, "Hypothalamus," In Wikipedia, no date.

48. No author, "Indusium Griseum," In Wikipedia, no date.

49. No author, "Septal Nuclei," In Wikipedia, no date.

50. No author, "Thalamus," In Wikipedia, no date.

51. No author, "Spinalthalamic Tract," In Wikipedia, no date.

52. Huang, Y., et al, "Measurements and models of electric fields in the in vivo human brain during transcranial electric stimulation," Retrieved at eLife 2017:6:e18834. See also: Irani, David, "Cerebralspinal Fluid in Clinical Practice," Chap.10, 69-89, Saunders, 2009. Agarwal, R.P., Henkin, R.I., "Zinc and copper in human cerebralspinal fluid," *Biological Trace Element Research*, 1982 June;4(2-3); 117-124; and Lipkova, J. and Cechak, J., "Human electromagnetic emission in the ELF band," Radar Dept., Faculty of Military Technology, University of Defense, Czech Republic, in *Measurement Science Review*, Volume 5, Section 2, 2005.

53. Understanding the stress response. 2011, *Harvard Health Publications*, Harvard University, Retrieved at health.harvard.edu.and "The Brain from Top to Bottom" (n.d.), McGill University, Canada, Retrieved at thebrain.mcgill.ca.

54. id.

55. Sargis, Robert, MD, PhD, "An overview of the hypothalamus: the endocrine system's link to the nervous system," 2015, endocrineweb, Retrieved at endocrineweb.com.

56. Harvard, supra.

57. Cortisol Initiates. (n.d.), In Wikipedia.

58. id.

59. id.

60. id.

61. Mashour, G., Borjigin, J. and Lee, U., "Surge of neurophysiological coherence and connectivity in the dying brain," 2013, PNAS, Retrieved at pnas.org. Also See Noble, PE., et al, "Tracing the dynamics of gene transcripts after organismal death,"

Open Biology, 2017, Jan. 7 (1), 160267, Retrieved at rsob.royalsocietypublishing.org.

62. Mastin, L., "The human memory," 2010, Retrieved at humanmemory.net.

63. id.

64. id.

65. id.

66. id.

67. Becker, J.S. and Zoriy, M.V., et al, "Imaging of copper, zinc, and other elements in thin section of human brain samples (hippocampus) by laser ablation inductivity coupled plasma mass spectrometry," *Analytical Chemistry*, 2005, May 15; 77(10): 3208-16, Retrieved at juser.fz-juelich.de.

68. Angelova, M. and Asenova, S., et al, "Copper in the human organism," *Trakia Journal of Sciences*, Volume 9, No. 1, 88-98, 2011.

69. Sofronescu, A., PhD, "Cerebral spinal fluid analysis," 2015, Medscape, Retrieved at emedicine.medscape.com.

70. Cerebralspinal fluid. (n.d.), In Wikipedia; See also, Klarica, M. and Oreskovic, D., "A new look at cerebralspinal fluid movement," 2014, Fluids Barriers CNS.

71. Bauman, S. and Wozny, D., et al, "The electrical conductivity of human cerebralspinal fluid at body temperature," 1997, *IEEE Transactions on Biomedical Engineering*, 220-223, Retrieved at ieeexplore.ieee.org.

72. "Ion Channels," (n.d.), Biology Reference, Retrieved at biologyreference.com; See also Ion Channel. (n.d.) In Wikipedia.

73. Dolcos, F., Lordan, A., and Dolcos, S., "Neural correlates of emotion-cognition interactions: A review of evidence from brain-imaging investigations," 2011, *Journal Cognition Psychology*, 669-694.

74. Kennsinger, E. and Lorkin, S., "Effect of negative emotional content on working memory and long-term memory," *Emotion, 2003 Dec;3(4), 378-93*.

75. Amaral, D. and Scharfman, H.E., et al, "The dentate gyrus: Fundamental neuroanatomical organization (dentate gyrus for dummies)," 2007, Progressive Brain Research, Retrieved at

ncbi.nlm.nih.gov.

76. "The puzzling role of biophotons in the brain," 2010, *MIT Technology Review*, Retrieved at technologyreview.com.

77. Rahnama, M. and Tuszynski, J., et al, "Emission of mitochondrial biophotons and their effect on electrical activity of membrane via microtubules," *J. Intergr. Neuroscience*, Volume 10, No. 1 65-88, 2011, Retrieved at arxiv.org.

78. id.

79. id.

80. Caswell, J. and Dotta, B.T., et al, "Cerebral biophoton emission as a potential factor in non-local human-machine interaction," *NeuroQuantology*, March, 2014, pp 1-11, Retrieved at neuroquantology.com.

81. Costa, J., Rouleau, N. and Persinger, M., "Differential spontaneous photon emissions from cerebral hemispheres of fixed human brains: Asymmetric coupling to geomagnetic activity and potentials for examining post-mortem intrinsic photon information," 2016, *Neuroscience and Medicine*, 49-59, Retrieved at scirp.org.

82. Rouleau, N. and Murugan, N.J., et al, "When is the brain dead? Living-like electrophysiological responses and photon emissions

from applications of neurotransmitters in fixed post-mortem human brains," 2016, *PLOS One*, Retrieved at journals.plos.org.

83. Wang, Z. and Wang, N, et al, "Human high-intelligence is involved in red-shift of biophotonic activities in the brain," 2016, PNAS, Retrieved at pnas.org.

84. Furusawa, A. and Sorenson, J.L., et al, "Unconditional quantum teleportation," *Science,* Oct., 1998, Volume 282, pp. 706-709, Retrieved at sciencemag.org.

85. Sun, Qi-chao and Mao, Ya-Li, et al, "Quantum teleportation with independent sources and prior entanglement distribution over a network," 2016, *Nature Photonics*, 671-675.

86. Valivarthi, R. and Puigibert, M.G., et al, "Quantum teleportation across a metropolitan fibre network," 2016, *Nature Photonics*, 676-680.

87. Einstein, A., Podolsky, B. and Rosen, N., "Can quantum-mechanical description of physical reality be considered complete," 1935, *Physics Review*, 777, Retrieved at journals.aps.org.

88. id.

89. id.

90. Photon. (n.d.), In Wikipedia.

91. Feynman, R., "Relation of wave and particle viewpoints," 1965, *Feynman Lectures, Vol. III, Quantum Mechanics, Chapter 2*, Caltech.

92. Balaguru, S. and Uppal, R., et al, "Investigation of the spinal cord as a natural receptor antenna for incident em waves and possible impact on the central nervous system," 2012, *Electromagnetic Biology and Medicine*, Retrieved at ncbi.nlm.nih.gov; See Also "Spine antenna pointing mechanism (APM)" (n.d.), MMA Design, Retrieved at mmadesignlic.com., MMA Design manufactures and sells a patent-pending antenna for space satellite-Earth communication. It mimics the human spine, using a stack of ten vertebrae-like metallic, pivoting elements, one on top of the other, with metallic coils serving as discs. The spine has a hole in the center running longitudinally through the coils (just like a human spine has a hole for the spinal cord). Instead of a spinal cord, a flexible RF coaxial cable or wave guide (analogous to the spinal cord) is routed through the center of the device. The purpose of this spine-shaped

antenna is to transmit and receive signals between spacecraft and ground stations during space missions.

93. id.

94. "Pregnancy: Week by week. (n.d.), Mayo Clinic, Retrieved at mayoclinic.org.

95. Median section embryonic brain, (n.d.), In Wikimedia.

96. Eisaman, MD and Fan, J., et al, "Single-photon sources and detectors," *Review of Scientific Instruments*, Vol. 82, 071101, 2011. Their article provides an excellent overview of the many

different types of single-photon detectors and their application.

97. Marsili, F., et al, "Single-photon detectors based on ultra-narrow superconducting nanowires," *Nano Letters*, May, Vol. 11, 2048-53, 2011.

98. Baek, B, et al, "Superconducting a-WxSi1-X nanowire single-photon detector with saturated internal quantum efficiency from visible to 1850 nm," *Applied Physics Letters*, Vol. 98,

251105, 2011.

99. Chunnilall, Christopher, J. and Pietro, I., et al, "Metrology of single-photon sources and detectors: a review," *Optical Engineering*, Vol. 53, 081910, 2014.

100. Sagan, Carl, *The Demon-Haunted World: Science as a Candle in the Dark*, Headline, London, 1997.

ILLUSTRATIONS

Illustration 1..Page 91

    Courtesy Wikipedia.com "Anatomy-Nervous System"

Illustration 2..Page 92

    Creative Commons Share-Alike 3.0 License courtesy Caio Arroyo 2010

Illustration 3..Page 98

    Creative Commons Share-Alike 3.0 License courtesy Oxford University Medical School and dansmedschoolnotes.com

Illustration 4..Page 108

    Creative Commons Share-Alike 3.0 License courtesy Tangent LLC 2018

Illustration 5..Page 115

    Courtesy Freefoto by Ian Britton (no date)

Illustration 6..Page 122

    Creative Commons Share-Alike 3.0 License courtesy regalrealness.com 2018

Illustration 7..Page 124

    Creative Commons Share-Alike 3.0 License courtesy Wikipedia.com

Illustration 8..Page 133

    Creative Commons Share-Alike 3.0 License courtesy Fatima Al Sayed 2013

Illustration 9..Page 150

    Pinterest 2018

Illustration 10..Page 152

Creative Commons Share-Alike 3.0 License courtesy sharonapbiotaxonomy via wikipedia.com

Illustration 11..Page 155

Creative Commons Share-Alike 3.0 License courtesy Wikipedia.com

Illustration 12..Page 155

wikipedia.com "136-174MHz base station antenna" courtesy David Jordan

Illustration No. 13..Page 155

Creative Commons Share-Alike 3.0 License

Illustration 14... Page 156

Courtesy Staff Sgt. Matthew Smith; 4th Combat Camera Squadron; MacDill AFB, Tampa

"Satellite Voice Antenna Installation—Operation Patriot Sands," 2013

THE AUTHOR

Michael A. Tewell worked briefly as a professional actor; a radio disk jockey, radio news director; and as a newspaper reporter—all before attending Kent State University at the age of twenty-five. At Kent State University, he received his Master of Arts in International Relations in 1979 and wrote his thesis on international terrorism (*Ideological Terrorism: An Analysis of the Capability and Will of Transnational Terrorist Groups to Attack the United* States, 1979).

He was admitted into Kent State Graduate College after his sophomore year and is one of the few college students to ever receive a master's degree without receiving a bachelor's degree. Tewell earned his Juris Doctorate degree from The National Law Center at The George Washington University in Washington, D.C. in 1982.

Rather that stay in Washington, D.C to work in national politics or the federal government, Tewell chose a legal career ensuring more direct contact with people's lives. He served twenty-five years as a trial lawyer defending the poor, the mentally ill, juveniles and veterans at the Office of Public Defender in Florida (Sixth Judicial Circuit).

In more than 30 years of practicing law, Tewell tried over 140 jury trials including homicides, manslaughter and other serious felonies and misdemeanors. He assisted in many other jury trials in a mentoring role and organized a trial practice clinic to teach new defense lawyers trial preparation techniques, pretrial motion practice, trial tactics and trial legal procedure needed to effectively represent clients at trial.

Since his retirement, Tewell enjoys life with his wife (also a criminal defense lawyer), their teenaged son, and a precocious cockapoo. His spare time is dedicated to writing on varied topics of interest, including spirituality, law, science and politics.

www.ingramcontent.com/pod-product-compliance
Lightning Source LLC
Chambersburg PA
CBHW071911290426
44110CB00013B/1347